Public Participation Process in Urban Planning

This book critically examines the public participation processes in urban planning and development by evaluating the operations of Planning Advisory Committees (PACs) through two meta-criteria of fairness and effectiveness.

Traditional models of public participation in planning have long been criticized for separating planners from the public. This book proposes a novel conceptual model to address the gaps in existing practices in order to encourage greater public involvement in planning decisions and policymaking. It assesses the application of the evaluative framework for PACs as a new approach to public participation evaluation in urban planning. With a case study focused on the PACs in Inner City area of Canberra, Australia, the book offers a conceptual framework for evaluating fairness and effectiveness of the public participation processes that can also be extended to other countries such as the United States, the United Kingdom, New Zealand, Canada, Scandinavian countries, the European Union and some Asian countries such as India.

Offering valuable insights on how operational processes of PACs can be re-configured, this book will be a useful guide for students and academics of planning and public policy analysis, as well as the planning professionals in both developed and developing countries.

Kamal Uddin, a Town Planner by profession, has been working in academia and the public and private sectors for more than 20 years in Australia, Bangladesh and Thailand. His interests are in line with the broader traditional forms of town planning, urban design, strategic management, community consultation and environmental design.

Bhuiyan Monwar Alam is a professor in the Department of Geography and Planning at the University of Toledo, Ohio, USA. He has 30 years of work experience in Bangladesh, Thailand and the United States. His major research interests are in transportation planning and environmental planning.

Routledge Studies in Urbanism and the City

For more information about this series, please visit https://www.routledge.com/Routledge-Studies-in-Urbanism-and-the-City/book-series/RSUC

Public Participation Process in Urban Planning

Evaluation Approaches of Fairness and
Effectiveness Criteria of Planning
Advisory Committees

Kamal Uddin
Bhuiyan Monwar Alam

Routledge
Taylor & Francis Group

LONDON AND NEW YORK

First published 2022
by Routledge
2 Park Square, Milton Park, Abingdon, Oxon OX14 4RN

and by Routledge
605 Third Avenue, New York, NY 10158

Routledge is an imprint of the Taylor & Francis Group, an informa business

British Library Cataloguing-in-Publication Data
A catalogue record for this book is available from the British Library

Library of Congress Cataloging-in-Publication Data
A catalog record has been requested for this book

ISBN: 978-0-367-64088-0 (hbk)
ISBN: 978-0-367-64089-7 (pbk)
ISBN: 978-1-003-12211-1 (ebk)

DOI: 10.4324/9781003122111

Typeset in Bembo
by KnowledgeWorks Global Ltd.

Contents

Figures

Tables

Acknowledgments

This book would have not been completed without guidance and support from quite a few people. Thanks are due to Professor George Cho and Professor Ken Taylor at the University of Canberra, Australia for their advice throughout the PhD research of the first author. This book is based on the main contents of the PhD thesis, but significant changes have been made by us to reflect the current situation in the ACT.

A special thanks to all Local Area Planning Advisory Committee (LAPAC) Members, Conveners, Coordinators, and Technical Officers of the Planning Authority who participated in the research project and spent long hours to discuss on planning and development matters in the Australian Capital Territory (ACT). Without their cooperation and comments on the various aspects of planning system, this book would never have been completed. Thanks are due to planning staff of the ACT Government for providing planning documents and information on the planning committees. We also wish to thank the team at KnowledgeWorks Global Limited with copy-editing the book.

Finally, we are indebted to our families, particularly Kamal Uddin is grateful to his wife Shahana and his wonderful children Subah and Shafin for their sacrifice.

Thank you!

Kamal Uddin and Bhuiyan Monwar Alam

Preface

Traditional models of public participation in planning have been criticized as 'top-down' approaches, segregating planners from ordinary citizens. Therefore, there has been a quest for decades for greater public involvement in planning decisions and policy making. The public demands a greater voice in planning and development affairs. To provide public input into the planning process, planning agencies often establish Citizen Advisory Committees (CAC) to involve the public in planning decisions. The increasing redevelopment pressure in inner-city suburbs in most Australian cities has led to the creation of many CACs for advising planning agencies and Ministers for Planning on planning and development matters. These CACs usually consist of people of diverse backgrounds who are elected, selected and/or appointed by the planning agency to provide community input into planning policy making.

However, little is known about the context and operational process in the consultation processes of PACs. Much of the existing literature on public participation lacks widely applicable evaluation approaches for determining whether the existing context and process is fair and effective in the participation process. As complex social phenomena, public participation processes are influenced by contextual factors. This book examines the Terms of Reference and the operational process of PACs and evaluates them through two proposed meta-criteria: fairness and effectiveness. Using a case study approach, this study selected PACs known as Local Area Planning Advisory Committees (LAPAC) in Canberra, Australia for evaluating public participation process, which provided a basis to develop a conceptual model for its improvement. The analysis was based on a theoretical framework, which focused on the criteria of fairness and effectiveness in the public participation process.

The book has used a qualitative approach to data analysis using multi-method techniques such as focus interviews, document analysis and participant observation. The interviews were conducted with LAPAC members and concerned planning communities, who were directly or indirectly involved in the ACT's consultation process and thereby aware of its planning decisions. They were development proponents, the enthusiastic wider public, planning staff, the Minister for Planning and planning spokespersons of political parties.

The data provided insights into the details of the proposed criteria to evaluate the fairness and effectiveness of a participation process. The results suggest that improving the participation process in a PAC requires changes in Committee protocols, operational processes, and planners' roles in conducting participation process. Specifically, there needs to be a move away from static processes toward more strategic, active and accountable processes. The book suggests some practical steps in order to ensure greater fairness and effectiveness in the participation process of a PAC, and recommends the proposed evaluative criteria as a new framework for evaluating PACs.

Keywords: planning advisory committees, public participation, fairness and effectiveness, qualitative approach, cast study approach.

1 Introduction

1.1 Background

The concept of public participation is not new to the planning community (Campbell and Marshall, 2000: 322); it was formally mandated in the early 1960s for urban renewal projects (Day, 1997: 423). For most renewal projects, public participation meant the creation of a seven-to-fifteen member advisory board mainly comprising community leaders: contractors, bankers, developers and legislators (Burke, 1979). The general public were not included at first, but in the late 1960s and the early 1970s, the city governments realized that the public should be involved by having a voice in and contributing substantially to the programs that affected their destiny, rather than being allowed to participate as a means to obtain their cooperation (Catanese, 1984). Friedman and Kaplan (1975) noted that in the early 1960s, the city governments were obliged to create a mechanism that allowed residents of the area to be involved in planning and carrying out the programs. Residents were to have opportunities to contribute ideas to the urban renewal programs and comment on its operation. The development proponents welcomed the involvement of residents, in the hope of increasing opportunities for the community to comment on development plans and government policies, and to identify issues that require attention by the government and planning agencies (Sewell and Phillips, 1979: 337). However, planning agencies emphasized improving communication between various citizen groups and planners, not adequately taking account of the community's views and their preferences in planning decisions (Arnstein, 1969; Day, 1997)

During the 1970s, many participation models were used to involve the community in the planning process but they were criticized as 'top-down' and 'technocratic' for not allowing a fair representation of public interest in decision-making (Arnstein, 1969). Such criticisms included a lack of fairness in the public participation process, which led to ineffective participation (Webler, 1995: 79). As a result, there were many renewed efforts to use various participation techniques to overcome these deficiencies and to provide a less confrontational and more integrative way to involve the public in planning decisions (Smith, 1993; Dickinson, 1999). Among these mechanisms

DOI: 10.4324/9781003122111-1

were advisory committees, which were established by the authority to provide advice on planning and development matters in the hope of channelling community contributions to the planning decision-making process (Thomas, 1995; Carson, 1996; Harding, 1998; Dickinson, 1999).

However, there is a lack of systematic evaluation by planning agencies of whether the process meets the participants' expectations of fairness and effectiveness. Webler and Tuler state, 'fairness refers to what people are permitted to do in the consultation process, when participants come together with the intention of reaching understanding and making public decisions in a fair process' (2000: 569). On the other hand, according to Canadian Council of Forest Ministers (1997: 113), 'effectiveness refers to decision which incorporates and mediates the broad spectrum of concerns on a given issue'. Webler (1992) notes that people do not participate if the process is not fair, and planners see no sense in participation that produces ineffective outcomes, so an analysis of public participation should focus on these two goals (Webler, 1995). This book attempts to evaluate the public participation process and develop a conceptual model that is fair and effective for a planning advisory committee. Using a case study approach, the author evaluate a type of planning advisory committee used to be known as Local Area Planning Advisory Committee (LAPAC). This advisory committee used to advise on planning and development matters to planning authority of the Government of Australian Capital Territory (ACT). In the recent past, the ACT Government changed the LAPAC into a different format for public participation in local area planning. This book uses LAPAC (hereafter we refer LAPAC as 'Committee') as a case study to evaluate the fairness and effectiveness in public participation.

1.2 Existing situation and the needs for this book

Since the United Nations (UN) Conference on Environment and Development (the Earth Summit), held in Rio de Janeiro in June 1992, governments in different countries have repeatedly professed their commitment to community consultation and greater participation in planning decision-making (James and Blamey, 1999a). As a result of this commitment, the government and planning agencies have been increasingly establishing environmental and planning advisory committees to ensure that decisions are integrated, responsive and comprehensive (Dickinson, 1999). There are currently many officially established opportunities for public participation in environmental decision-making in Australia (James, 1999). Taberner and Brunton (1996) noted that there were some 320 statutes, which dealt either principally or in part with environmental and planning matters in Australia. Most of the city councils across Australia have established advisory committees for environment and planning matters (Taberner and Brunton, 1996). These advisory committees, often known as Citizens' Advisory Committees (CACs), are usually composed of a group of people either elected or appointed by the councils to provide a mechanism for public comment and input in the planning decision-making process. Through this medium the community can contribute to processes such as government decision-making and council's planning policies and management practices (Harding, 1998).

However, the advisory committees have been criticized by definition as ineffective in the theoretical literature (Houghton, 1988) as not being open and fair forms of public participation, since 'selected members of a committee may be acting with vested interests in the decision or development proposal under review rather than in the interests of the wider community' (Harding, 1998: 116). Arnstein (1969) commented that 'in the name of citizen participation, people are placed on rubberstamp advisory committees or advisory boards for the express purpose of "educating" them or engineering their support, and that advisory boards may be formed by the proponents as a public relations exercise to "win" the support of neighborhood and community groups' (Arnstein, 1969: 218).

Given the criticisms of this form of public participation, 'the advisory committees are seen as an alternative to the "shallow and broad" participation of public hearing or survey'. They are deeply involved in a process, but ... 'a narrow segment of the population instead' (Webler, 1992: 222). On the other hand, CACs are the regular forms of consultation bodies where long-time participants on a committee become experts in explaining the issues, they discuss and take a great amount of commitment in time and effort. The long-time participants become familiar with the consultation process and problems that are on the agenda for the greater discussions and comments. Thus, a systematic evaluative approach is required to examine advisory committees in order to analyze their consultation process and effectiveness in planning decisions. The objective of the evaluation is to improve the existing consultation process and to form a better participation approach to be adopted by the planning agencies. The existing literature indicates that most of the evaluation case studies are on various issues of environmental decision-making (Beierle, 1998, 1999; Webler and Renn, 1995). CACs on environmental decision-making are mainly composed of technical experts and rarely include laypersons on the committee. Planning advisory committees (PAC) on the other hand, consist of people with diverse interests and include professional and non-professional laypersons, residents, community groups and interested development proponents. There is a lack of available literature on planning advisory committees and their roles in planning decisions. This book will contribute to the planning literature by investigating the consultation process from a case study perspective. Using focus interview techniques, participant observation and document analysis, this book develops a conceptual model that addresses the fairness and effectiveness of the participation process in planning advisory committees.

In the ACT, participation used to occur in various forms including public hearings, opinion polls, consensus conferences, deliberative democracy and advisory committees. There were advisory committees in the ACT that used to advice on many issues of interest to the ACT Government and its various departments. Some special departments such as the ACT Planning and Land Authority (ACTPLA) formed advisory committees in 1995 to involve community, residents and businesspeople in planning decisions. These advisory committees meet regularly, but final decisions are made within the planning authority's decision-making process. However, members of the committees have had reported that they had very little to say about their opinions. Sometimes they are not well informed about the continuing process of any development activities and often their recommendations are not incorporated in the planning decisions. At the same time, the planning advisory committees are concerned about the recruiting of members to the committee, which they believe is undemocratic. People interested in participating in the continuing activities are selected for the planning advisory committees and serve objectives of the planning authority. The author of this book attended several meetings and was informed by the planning advisory committee members that the recruiting policy was not fair and often served only the planning authority's objectives.

To date, the planning advisory committee has attracted very little attention in the planning literature, and its effects on local planning decisions have not been systematically evaluated. Day (1997) notes that there was a recession in planning research in the early 1980s when interest in public participation was sidelined, and planners instead focused on issues of strategic planning, economic development and environmental decision-making. Thus, much of the public participation literature tends to concentrate on environmental decision-making and the issues of environmental impact assessments (Grant, 1994). Although concerns with public participation in planning did not disappear, it was given less consideration than the environmental issues (Day, 1997). However, some of the environmental advisory committees, often called environmental advisory groups (EAGs), also deal with urban planning issues at a policy level that integrates planning and environmental issues. Hence, evaluations of planning advisory committees become inevitable in order to understand the advantages and disadvantages of the planning advisory committees, its contexts and the process of consultation.

This book is a critical review of the evaluation approaches of public participation in planning. It suggests the needs to evaluate the broader contexts of the planning advisory committee and its operational processes. The book also develops a theoretical framework for evaluating public participation process in planning advisory committees. It discusses the existing evaluation criteria while proposing two meta-criteria: fairness and effectiveness in evaluating public participation process. The evaluation of the planning advisory committees consequently leads to developing a conceptual model to improve existing practices. The book is expected to influence the planning agencies in designing future planning advisory committees. Additionally, the book fills some major gaps in the theoretical literature, such as the criteria for evaluating the fairness and effectiveness of public participation in planning in general and planning advisory committees in particular.

1.3 Objectives of the book

The main objective of the book is to critically evaluate a planning advisory committee through the proposed meta-criteria of fairness and effectiveness, and to develop a conceptual model that is fair and effective. The sub-objectives are to:

1 examine the existing models of evaluating public participation in planning
2 evaluate planning advisory committees through the proposed meta-criteria of fairness and effectiveness
3 develop a conceptual model in order to address the existing practices in public participation and
4 assess the application of the evaluative framework for planning advisory committees as a new approach to public participation evaluation in planning literature.

1.4 Significance of the book

This book makes a conceptual contribution to the development of an evaluative framework to examine public participation process in planning through planning advisory committees. To date, there has been no method available for evaluating fairness and effectiveness in public participation in planning The literature search also suggests that there has been no study to evaluate whether the participants of the planning advisory committees thought that the exercise was conducted in a fair and effective way, nor to compare the effectiveness with different contexts and situations. This book provides guidelines for the planning agencies to evaluate the existing public participation processes.

Much public participation research has focused on evaluation of particular involvement techniques (Heberlein, 1976; Checkoway, 1981; Priscolli, 1983); however, no single public participation mechanism appears as the best across all situations (Thomas, 1990). As a complex social phenomenon, public participation is affected by the local context, which may relate to a planning agency and its process, the community and community-based organizations, and their expectations from the consultation process. Considering the local context in ACT, the planning advisory committees are more focused on planning and development matters, raising them with the planning agencies on a regular basis. Hence, the book evaluates CAC techniques by considering planning advisory committee as a case study. The book also evaluates the consultation process with the help of a framework. The significance of the book is enormous in that it is the first book that evaluates planning advisory committees using meta-criteria of fairness and effectiveness, and develops a conceptual model for public participation that is fair and effective.

2 What we know about fairness and effectiveness in public participation in urban planning

2.1 Introduction

This chapter provides a review of the literature on the existing models of public participation in planning. The review ranges from methods of public participation in planning and approaches to the participation using examples from the available literature and case studies. The review also includes two sections. The first section discusses the definitions of participation and their origin in the planning literature and methods of public participation. The second section discusses the role of advisory committees in decision-making, and the policy community model as a basis to evaluate the public participation process in planning and development matters.

Public participation is a broad concept and has received widespread attention in both empirical and theoretical research (Day, 1997; Innes and Booher, 2000; Lauber and Knuth, 2000; Palerm, 2000). It has been a major focus of research in both human-dimensions disciplines like biology, environment, natural resources and waste management and social science research such as planning, geography, sociology and political science (Lauber and Knuth, 2000). This chapter presents a critical review of public participation in planning theories and their influence on planning decisions. It also reviews the 'policy community model' (Pross, 1992), which is applied to municipal environmental advisory groups (Filyk and Cote, 1992; Dickinson, 1999). This model adapts to the 'urban planning policy community' as a basis for evaluating urban planning advisory committees and their role in planning decisions.

2.2 The concept of 'participation' and its origins in the planning literature

Day (1997: 422) argues that the theoretical literature on public participation in planning process is as cumbersome as that of empirical literature. The existing planning literature is also inundated with many definitions of participation. The widespread use of the term 'participation' has tended to mean that any precise, meaningful content has almost disappeared (Pateman, 1970), because 'participation' is used by different people to refer to a wide variety

DOI: 10.4324/9781003122111-2

of situations (Burke, 1979). In addition, there are many adjectives before the word 'participation', as in 'community participation', 'citizen participation', 'people's participation', 'public participation', 'public involvement' and 'popular participation' (Mathbor, 1999: 14). These terms are used interchangeably, but in the planning literature the three terms 'community participation', 'public participation' and 'citizen participation' are commonly used. Public participation and citizen participation are mostly popular with the planners, and community participation is mostly used in development literature (Mathbor, 1999: 14).

However, there are differences in meaning between *public participation* and *citizen participation*. Public participation is an act of taking part in the formulation, implementation and evaluation of policies by interest groups through formal institutions (Yigitcanlar, 2000). Examples of interest groups include many professional and community groups such as residential associations, business associations, trade unions, professional advisory groups and staff associations. Citizen participation, on the other hand, is the direct participation of ordinary citizens in public matters. Citizen participation, however, is distinct from political participation, because citizen participation lays emphasis on the person rather that the state in the participatory relationship. Public participation is not synonymous with citizen participation, mainly because public participation is a wider concept, which may include citizen participation (Yigitcanlar, 2000). This is due to the fact that the word 'public' refers to all people whether or not they possess the rights and obligations of citizenship. Public participation is taken to include citizen participation. Yigitcanlar (2000) notes that public participation in planning is an approach in which citizens are brought to play an active role in the planning decisions.

As mentioned above, 'participation' is a contested concept (Day, 1997), so the meaning of participation varies in relation to its applications and definitions. The way that participation is defined also depends on the context in which it occurs. For some, it is a matter of *principles* (Pimbert and Pretty, 1997) for some others, it is *practices* (Davis, 1996); and for still others, it is an *end* (Rahman, 1993). However, there is merit in all these interpretations. Rahnema (1992: 116) notes that 'participation is a stereotype word like children use Lego pieces. Like Lego pieces the words fit arbitrarily together and support the most fanciful constructions. They have no content, but do serve a function. As these words are separate from any context, they are ideal for manipulative purposes'. Similarly, Arnstein (1969: 216) argues that 'participation is a little like spinach: no one is against it in principle because it is good for you'.

Schatzow (1977: 141) defined participation as 'something that is distinguished from public influence', asserting that participation refers to the involvement of the public in decision-making through a series of formal and informal mechanisms. Day (1997: 493) observes that there are differences between public participation and public influence. For him, public participation and decision making do not guarantee that the public is able to exert their opinion and influence. Instead, public influence is attributed to the situation when public's opinion is reflected in the decision-making process and actual decisions made (Schatzow, 1977: 142). On the other hand, Arnstein (1969: 216) defined participation as a categorical term for citizen power. For Arnstein (1969: 216), participation is

> ...the distribution of power that enables the have-not citizens, currently excluded from political and economic processes, to be deliberately included in future. It is the strategy by which *have-nots* join in determining how information is shared, goals and policies are set, resources are allocated, programs are operated, and benefits like contracts and patronage are parcelled out.

Arnstein (1969: 216) continues to observe that participation is the means which the public can utilize to persuade social reforms that they can use to empower them to reap the benefits of the affluent society.

Therefore, 'participation' is seen as a way of influencing decisions that affect the lives of citizens, and to some extent, a way of transferring political power to the grassroot people (Brager et al., 1987). Chambers (1997) sees participation as a process of involving affected communities and interested individuals to participate in decision-making process, in response to community concerns, have their voices heard, and bear the obligations for changes within their communities (Chambers 1997: 68). Similarly, the World Bank's Learning Group on Participatory Development defines participation as a means by which the interested parties voice their opinion and exert influence in making decisions over development activities that directly or indirectly affect them (World Bank, 1995: 3). Slocum and Thomas-Slatyer (1995: 113) and Harding (1998: 108) define participation as an act by which the affected people engage themselves in the decision-making and implementation process of projects and programs that affect their livelihood. The central point of this definition is the belief that ordinary people are capable of producing critical reflections and their place-based local knowledge is relevant and necessary for the decision-making process.

Whilst many use 'public participation' to cover a range of interactions between two typologies namely 'consultation' and 'empowerment', there are differences in the meanings of both typologies (Buchy and Hoverman, 1999). 'Consultation' is presented as a process of involvement in which people's opinion is sought, which may influence the perspective, but which in no way guarantees an input in decision-making (Buchy and Hoverman 1999: 9). It has also been described as 'sharing of information but not necessarily of power' (Sarkissian et al., 1997). When people are consulted before the preparation of a new project, their opinion is more likely to be incorporated than if they are asked to comment on previously identified and designed project. On the other hand, 'empowerment' reflects more of a state of personal development and a state of mind through which people learn, increase their self-esteem and confidence, and are better able to use their own resources (Chambers, 1997).

Public participation is required in order to ensure that the aspirations and needs of citizens are taken into account by the decision-makers, as more people and communities emerge with the demand for their voices to be heard and included in the decision-making process (Harding, 1998: 113). The modern educational and democratic norms are very important in making people aware of their rights to participate in a program that may directly or indirectly affect them. There are numerous models to involve the public in the planning process. Since the 1950s, people have become involved in planning decisions, particularly in urban renewal projects (Day, 1997). Governments and their planning agencies have introduced various techniques and methods to involve the possibly affected communities and enthusiastic individuals in the decision-making process. The most popular techniques have appeared to be public hearings and formation of advisory groups consisting of selected individuals in the communities (Day, 1997: 495). Innes and Booher (2000: 1) criticized these methods by claiming that they do not achieve genuine participation in planning or decisions, do not provide significant information to public officials that makes a difference to their actions, do not satisfy members of the public that they are being heard, do not improve the decisions that agencies and public officials make, and do not represent a broad spectrum of the public. Birkeland (1999: 113) observes that urban planning is frequently depicted as a top-down approach taken by the technocrat planners, while the solutions are sought in bottom-up community participation approach. Innes and Booher (2000) also advocate a 'bottom-up' approach within the framework of the collaborative planning process.

The planning literature indicates that there is a potential to create a collaborative approach to planning decisions by planning stakeholders, which will require a collaborative participatory approach by the planners to the execution of their duties and to the needs of society (Gray and Wood, 1991; Sandercock, 1998; Birkeland, 1999; Innes and Booher, 1999a, 1999b, 2000). The public should realize that individually or through interest groups, they should participate in public matters that may affect them, with a view to persuading planning agencies to promote their particular interests by changing public policy on specific matters. Everyone also has to realize that public participation can shape the broad policies of local governments, and its effects on routine policy decisions could be maximal (Sandercock, 1998). The following Section 2.3 discusses theories of public participation and the main features of public participation in collaborative planning.

2.3 Public participation in planning theories

Over the last several decades, many competing models of planning theories for public participation have been developed (Sandercock, 1998; Birkeland, 1999; Innes and Booher, 1999a, 1999b, 2000). Many of the planning theories

	Low	Diversity	High
	Low	**Technical/Bureaucratic** Convening	**Political Influence** Co-opting
Interdependence of Interest			
	High	**Social movement** Converting	**Collaborative** Co-evolving

Figure 2.1 Four models of planning and policymaking.

Source: Innes and Booher (1999b)

on which the participation models are based overlap, and all continue to be used in present-day planning practices (Sandercock, 1998). Innes and Booher (1999b) identify four schools of planning thought and explore how each school approaches the issues of public participation. These schools are: technocratic, political, social movement, and collaborative participatory planning (Figure 2.1).

Another list of public participation models under various planning theories was summarized by Birkeland (1999), and contains four competing models of public participations in planning process. These include (1) technocratic, (2) liberal, (3) radical and (4) eco-centric bioregional through collaborative approach (Table 2.1). However, this book summarizes planning theories under four headings based on the nature and manifestation of planning theories in the participation process, viz., comprehensive planning; incremental planning; advocacy planning and collaborative planning.

2.3.1 Participation in comprehensive planning

Public participation in comprehensive planning approach is a version of the 'techno-bureaucratic' or 'top-down' model (Birkeland, 1999; Innes and Booher, 1999b). In this model, planning is about assessing the alternatives that best meet goals, developing comparative analysis, recommending to decision-makers a course of action and, later assessing the effects of policies and suggesting possible changes (Innes and Booher, 2000). This planning process typically depends on scientific information and quantitative data modelling. Planners believe that there is a truth 'out there' which they are the

Table 2.1 Summary of participatory planning models

Planning theories	Participation models	Approaches
Comprehensive	Technocratic	Top-down
Incremental	Liberal	Non-planning
Advocacy	Radical	Bottom-up
Collaborative	Eco-centric Bioregional	Collaborative

Source: Adapted from Birkeland (1999: 114)

best qualified to uncover through their analytical skills. Public participation in this model is regarded by the planners as something to meet the obligations of planning laws (Sandercock, 1998; Birkeland, 1999; Kaufman, 1999).

Therefore, participation in this model means 'consultation' or 'input' to planning and development approval systems, while planners or technical experts determine the planning decisions on what is best for the possibly affected communities (Sandercock, 1998). Comprehensive planning assumes that an optimal result for the community can be objectively determined and that planning decisions flow directly from information. However, planners in comprehensive planning may also regard consultation as a useful way to better understand community goals and values, and to fill gaps in what has been formally adopted by Legislative Assembly for its objectives (Innes and Booher, 2000; Sandercock, 1998). Some planners may see the consultation process as wastage of time, and believe that the community has little to contribute to their highly acclaimed planning work which often comes with high-quality architectural designs and plans. In some cases, such technical planners may take local knowledge from the community, but look with much skepticism on 'anecdotal' evidence that comes from the public (Webler, 1995; Sandercock, 1998). The technical planners consider that public participation is the activity that is needed at two stages of the planning process – at the beginning stage to ponder and determine objectives and goals, and towards the end to select the best alternative as the final choice of the plan or strategy (Innes and Booher, 2000: 15).

2.3.2 Public participation in incremental planning

By the end of the 1960s, the limitations of comprehensive planning were evident (Gunton, 1984). A key element came from the Chicago Housing Authority by Banfield (1961) who argued that planning practice was different from the theory of planning. The theory of comprehensive planning was not a rational activity governed by experts using scientific knowledge, but an irrational process dominated by petty political concerns (Gunton, 1984). Arguing that decisions are sometimes taken by high-level officials instead of professional planners, Innes and Booher (1999b) call this approach 'political influence planning'. They note that, typically in this model, a plan is made up of projects, each of which is desired by a politically important person. For instance, a political leader or political party may select an extension route to connect the city with urban conurbation, which may benefit a political party in general, and a political leader in particular.

A vital element of public participation in incremental planning is a pluralistic view of a society composed of competing interest groups who lobby government for certain policies (Sandercock, 1998). In this model, planning decisions are not made by a strict process but by a series of consultations; thus decisions are made incrementally through a series of smaller decisions (Gunton, 1984), in order to lessen the danger of making big mistakes

(Birkeland, 1999: 121). This is the theory of Charles Lindblom (1959) who described it as 'partisan mutual adjustment' or 'disjointed incrementalism' (Faludi, 1973). Over time, incremental choices form a decision-tree: at each branch, the planning decisions may be rational, but taken as a whole they may not be, as people could end up out on a limb (Birkeland, 1999: 116).

In incremental model of planning, broad public participation is undesirable, if not actually a threat to the whole planning system (Sandercock, 1998). This type of planning works 'behind the scenes with the "fixer" making deals with powerful players one by one. It does not deal with the people who are not politically important to the political leaders; rather, it gives importance to the voters' (Innes and Booher, 2000: 14). Innes and Booher (1999b) note that this kind of deal-making does not bear public scrutiny, even when there is nothing illegal about it, since it violates the social norms of fairness and equal opportunity for all participants in the consultation process. Innes and Booher (1999b) also observe that public participation in this model often happens after the deals or preliminary decisions have been made. The public may then have the opportunity to comment on a decided proposal, though it is unlikely that basic changes will be made. Indeed, it is this deal-making component that is behind much of the impression the public often gets that participation is merely for window-dressing, as opposed to being designed to get greater community input into the planning decisions. Often the community regards this process as illegitimate and unfair because they believe the decision has already been made before the beginning of public consultation with the affected community. Sandercock (1998) notes that this model is the biggest obstacle to genuine public participation in decision-making process. Innes and Booher (2000: 16) argue that meaningful public participation can be achieved when incremental planning approach is coupled with technical approach

2.3.3 *Public participation in advocacy planning*

A third model of planning gives a voice to marginalized communities who often feel disenfranchised from participating in decision-making (Innes and Booher, 2000). Beginning with the 'War on Poverty' in the 1960s, some planners and architects began to realize that 'ghetto dwellers' had their distinct life styles (Wear, 1996). It was believed that these people should be mobilized by having a voice in and contributing substantively to programs that affected their destiny, rather than being allowed to participate as a means to obtain their cooperation (Burke, 1979). The aim of the 'War on Poverty' was to provide adequate civic amenities to poor slum-dwellers, providing an opportunity for advocacy planners and design agencies to incorporate people's perceptions of their lives and to improve their livelihood. Inspired by many social movements (Innes and Booher, 2000) to improve the lives of slum-dwellers during the 1960s, the advocacy planners sought to give these disadvantaged communities a voice in landuse planning which greatly affected their lives. This planning model is also known as bottom-up planning (Birkeland, 1999).

Public participation under this planning model means that advocacy planners try to empower a community by providing technical support and political advice, without imposing their own values, decisions or strategies on their client groups (Forester, 1999). Advocacy planners worked to overcome class, language and cultural barriers to assist under-represented community groups in communicating with technocrats and negotiating with administrators (Birkeland, 1999: 128). The essence of this model is that individuals and groups who are not in the power structure join together for some common purposes because the only way they can have an influence is through their number. In this process, the community takes an active role in planning and designs through hands-on involvement, rather than through 'consultation' (Sanoff, 2000). Many advocacy planners are parts of this process, whether formally or not, and pursue their mission in their planning activities, for non-government organizations or planning agencies or even in consulting practices. But often advocacy planners are volunteers working in their free time (Sanoff, 2000). Whether the advocacy planners work outside the planning agencies or inside government organizations, their objective is to improve participation or, at least, reduce the power differences between vested interest and community groups, rather than to change the decision-making systems fundamentally (Forester, 1999).

2.3.4 *Participation in collaborative planning*

Collaborative planning is a process in which different stakeholders with wide array of vested interests, work together for a common cause (Gay, 1989: 5). In the collaborative model, the essential idea is that planning should be done through face-to-face dialogue among those who have interests in the results. For this dialogue to work effectively, Innes and Booher (2000: 18) identify six pre-conditions for a participation process: (1) the full range of interests of the 'planning community' must be involved; (2) the dialogue must be authentic in that people must be able to speak sincerely and comprehensibly to each other; that what they say must be accurate and that they must speak as legitimate representatives of a stakeholder's interest; (3) there must be both diversity and interdependence among the collaborators; (4) all issues must be on the table for discussion with nothing off-limits; (5) everyone in the discussion must be equally informed, equally listened to and thus empowered as members of the collaborative discussion and (6) agreements are only reached when consensus is achieved among the vast majority of participants and only after substantial serious effort has been made to satisfy the interest of all those involved.

Innes and Booher (2000) note that the above conditions are closely related to the ideas of Susskind et al. (1999) and Habermas (1987). However, most of the pre-conditions have been discussed by Webler (1995) in his theory of fairness and competence in citizen participation in environmental decision-making. In this collaborative model, participants jointly develop their own objectives and purposes, structure their interests and preferences for all to understand, develop a shared understanding of a planning problem and agreement on what they need to do and then work through a series of tasks which lead to actions or agreements that all, or most, believe will improve their ability to meet their own interests and improve the collective welfare. The participants reach these results, not by argument, but by cooperative scenario-building, information and experience. All members have to create new strategies that can often release the group from some impasse that would otherwise have prevented action (Innes and Booher, 1999a). Innes and Booher (1999a) argue that traditional participation methods are not particularly satisfactory for many of their purposes, and point out that the collaborative planning process that comes close to meeting the ideal conditions incorporates participation in a much deeper, more inclusive and more meaningful sense than the conventional methods. Achieving these goals requires establishing a collaborative network paradigm with the planning stakeholders (Innes and Booher, 1999a). Innes and Booher (2000) also comment that the collaborative approach is the only method of planning and public involvement that would be sufficiently flexible, responsive and adaptive to be effective in the uncertain and rapidly changing environment of the turn of the twenty-first century (Innes and Booher, 2000: 14).

2.3.4.1 Approaches to public participation

Numerous authors have advocated public participation as a way to allow citizens to pursue their own interests, desires and preferences. Pateman (1970: 14) claims that public participation is helps protect various private interests. Therefore, the role of the government is to help citizens fulfil their interests and wishes (Fiorino, 1990).

Arnstein's (1969) 'A Ladder of Citizen Participation' is an oft-cited reference on participation in planning (Figure 2.2). It identifies eight different rungs of participation based on power in actual decision-making authority. Arnstein (1969: 217) coins citizen participation as the redistribution of power among those citizens who are considered as 'have-nots'. She continues to argue that participation is meaningless and frustrating if it is not redistributed among the general citizens.

The eight rungs are arranged with an increasing level of power. These eight categories of power fall into three main groups. The objective of the first group, namely the non-participation group, is actually to help enable the powerholders to 'educate' or 'cure' the general participants who belong to 'have-nots' category, and not to help them actively participate in planning activities or managing programs (Arnstein, 1969: 217). The group describes varying degrees of tokenism where the 'have-nots' are allowed to hear and to have a voice, but are not afforded the power to ensure their views are heard, nor given the right to decide (Arnstein, 1969: 217). With increasing citizen power, further up the

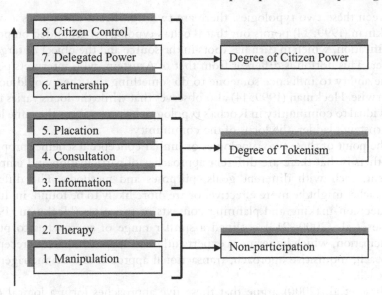

Figure 2.2 Eight rungs on a ladder of citizen participation.

Source: Arnstein (1969: 217)

ladder, the degree of decision-making power 'is transferred from the "power-holders" to the citizens, providing opportunities for negotiation, formulation of trade-offs and full managerial power' (Arnstein, 1969: 217).

(Byrne and Davis, 1998: 13) criticizes Arnstein's (1969: 216) ladder as arguing that it is judgmental in spreading the concept that some methods are inherently more valuable than others. It is also criticized as too theoretical and thereby idealistic, where one can hardly reach the top part of the ladder (Kaufman, 1999: 134). Twenty-eight years after Arnstein, Rocha (1997) presented another ladder of participation (Figure 2.3). While there are similarities

Community Empowerment	
Rung 5	Political Empowerment
Rung 4	Socio-political Empowerment
Rung 3	Mediated Empowerment
Rung 2	Embedded Individual Empowerment
Rung 1	Atomistic Individual Empowerment
Individual Empowerment	

Figure 2.3 A ladder of empowerment.

Source: Rocha (1997: 35)

between these two typologies, there are two significant differences as well. Heckman (1999: 14) points out that Rocha's typology is grounded on power classification, which in turn has roots in the source and the object or target of power. This notion is different from that of Arnstein, who interprets power as the ability to influence someone to do something that they would not do otherwise. Heckman (1999: 14) also observes that while the locus varies from individual to community in Rocha's typology, the rungs share the same locus in Arnstein's ladder, the locus of the community.

The point of these two typologies or similar conceptual schemes proposed by others is that there are different approaches of participation and empowerment, each with different goals, purposes and methods. The different approaches might be more effective, or are more likely to be found, in different decision-making and planning contexts or processes (Heckman, 1999). Chase et al. (2000: 213) outlined a similar range of approaches to public participation, which he terms as expert authority approach, passive-receptive approach, inquisitive approach, transactional approach and co-management approach.

Chase et al. (2000) argue that these five approaches form a logical continuum in which the relative influence of citizens and agencies on management varies, from total agency control under the expert authority approach to broad power-sharing under co-management.

2.3.4.2 Purposes of public participation

A number of purposes have been advanced for public participation in planning and policy decision-making. Glass (1979) identified five purposes of public participation in planning decisions: (1) information exchange among planning stakeholders; (2) education for the residents and community groups to make informed comments on planning proposals; (3) support building; (4) supplemental decision-making and (5) representational input in the decision-making process. Innes and Booher (2000: 5–6) outlined another list of four major purposes of the participation in planning, which is similar to that of Glass (1979).

The first is simply for decision makers to acquire information about community preferences so they can play a part in the decision-making process about projects, policies or plans. Here, public participation is designed to help ensure that the people's will is given consideration in planning decisions.

The second purpose is to improve the decisions that are made by incorporating the knowledge of the public into the realms of decision-making. Thus, it may be that the people in a local community know about the traffic or crime problems on a particular street, and the planners and decision-makers can learn about this through public involvement in decision-making process. Both the first and second purposes are increasingly important for planning agencies, as the agencies get larger and more impersonal and more distant from its constituencies.

The third purpose of public participation generally concerns fairness and justice. There are reasons why the disadvantaged groups' needs and preferences are likely to be unrecognized through the normal analytical procedures and information sources of bureaucrats, legislative officials and planners. These needs may only come to the planner's knowledge when an open public participation process occurs. Hence, public participation gives at least the opportunity to hear people who were overlooked or misunderstood in the early stages of planning.

A fourth purpose of public participation is about making the plans democratic and legitimate (Innes and Booher, 2000: 6). In many circumstances, public participation is used to build support for plans among the public. Conrad et al. (2011) observed that public participation is an essential element of most policies and strategies as a legal requirement within planning jurisdiction in many countries.

2.3.4.3 Public participation processes

The purposes of public participation discussed above have certain types of processes that contribute to the planning decisions (Figure 2.4). When

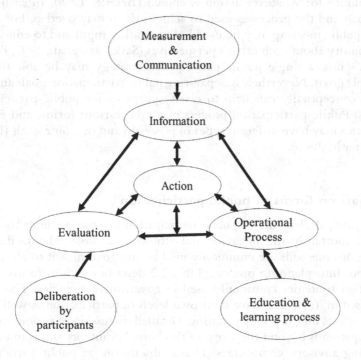

Figure 2.4 Relationship between public participation process and outcomes.

Source: Adapted from Lauber and Knuth (2000: 13)

planning agencies attempt to gather information about citizens to inform decision-makers, they rely on processes that involve measurement such as mail and telephone surveys, and soliciting feedback at public meetings (Lauber and Knuth, 2000). When the public try to provide information to decision-makers, they are faced with the task of effective communication to managers. When public participation is used to improve judgement, the public are incorporated in a process of deliberation, and discussion of the merits of different planning options. Lauber and Knuth (2000) note that if the management of operational process is the target of participation activities, some types of transformation of people is the purpose, whether a transformation of their perspectives, behaviors, relationships or capacity. Education of the relevant participants in the planning process can lead to personal transformation of individuals or groups. All these processes can contribute to decision-makers' willingness to implement planning actions (Lauber and Knuth, 2000). Thus, public participation is a continuous learning process for the participants in order to make an informed comment on the proposals.

In many public participation programs, multiple processes are used to achieve the goals. Planners may want to gather input, involve the public in deliberation for the best action to take, and build support within a community for whatever action is chosen (Beierle, 1998). In such cases, the goals and the processes used to achieve them may overlap. For example, a public meeting may be used both to gather input and to educate the community about each other's perspectives (Sarkissian et al., 1997; Beierle, 1998). Thus, a single public participation strategy may be able to meet multiple goals. Nevertheless, separating public participation goals and processes conceptually can help to clarify purposes for public participation efforts. Public participation process may take various forms, and each of the forms may have different types of processes and outcome goals (Lauber and Knuth, 2000).

2.4 Various forms of public participation

The principal elements of public participation are the methods used, and most importantly the behavior and attitudes of those who facilitate it. Numerous methods are commonly used by the government to allow public input into planning process. Table 2.2 describes various forms of participation strategies commonly used in government agencies. The forms, discussed in Table 2.2, have their own levels of participation as well as different effects on the policy planning. Detailed discussion of various forms of participation is beyond the scope of this book because its aim is to evaluate planning advisory committees. Hence, discussion on public participation has been confined to a particular technique, namely the Citizens' Advisory Committees (CAC).

Table 2.2 Public participation strategies

Forms of participation	Definition
Citizens' Advisory Committee (CAC)	People are selected by an institutional body based on representation of major interest positions, but not the full range of interests for logistical reasons.
Public hearing	People are invited to present their concerns before a lawful committee that may comprise planning officials and lawmakers.
Workshop	A process to gather community and stakeholders' input into a process that requires planning initiatives, which the agency thinks it important to incorporate.
Survey	Similar to workshop, the target people are identified to solicit their concerns about possible changes and satisfaction over the existing process and outcomes.
Citizen Taskforce	A process whereby a group is formed to devise equitable outcomes on planning issues to be considered
Planning Cell	People are selected from a random pool of citizens to evaluate
Citizen jury	People are selected from a random pool of the public to evaluate policy alternatives.
Citizen panel	A process that selects enthusiastic individuals to give policy level input into the decision-making process
Consensus conference	People are selected from among expert to make comments mostly on scientific and technological aspects.
Deliberative poll	People are selected randomly by telephone numbers and then come together to discuss the issues, thereby building in a deliberative component.
Public participation GIS	People are selected to shape, reshape and make alternative to their own areas of concern on planning and environmental aspects. It is also be possible through the Internet.

Source: Adapted from Jankowski and Nyerges (2001: 30)

2.4.1 Citizens' advisory committees

A CAC, which represents a particular community or neighborhood, is a form of participation body. Public participation can be ensured through a form of elected advisory committee comprising people from all walks of life. It may include residential, environmental and business groups, labor unions and agency staff as well as citizens' groups (Thomas, 1995). Thomas (1995) observes that an advisory committee is not a form of deliberative democracy but a type of 'republican' form of public involvement in participation, restricted to a small number of community representatives who are expected to represent the interests of the larger community. In the United States, the establishment of advisory committees sprang from initiatives such as the 'War on Poverty' during the 1960s and 1970s, to advice on various programs such as affordable housing for poor urban dwellers and, improving their lives. Advisory committees can also be formed to advise in the decision-making of governments or any organizations on planning issues. In Australia, political pressure by democratic interests from the 1970s helped to secure a number of reform initiatives in planning legislation and practice that sought to improve public participation mechanisms (Gleeson and Low, 2000). Stein (1998) notes that the New South Wales *Environmental Planning and Assessment Act 1979,* for example, enhanced the opportunities for public participation in that state's planning process. In the ACT, the *Planning and Development Act 2007* requires broad community involvement for development. In the Victoria, Advisory Committees are appointed by the Minister for Planning under Section 151 of the *Planning and Environment Act 1987* to consider development proposals or to review planning policies. Accordingly, many government agencies formed advisory committees to give advice and recommendations to the government's planning initiatives on many development projects. The advisory committee ranges from technical experts' panels to laypersons giving advice on many issues as community input. However, the literature indicates that most of the CACs are formed to deal with issues related to environmental decision-making (Lynn and Busenberg, 1995).

Lynn and Busenberg (1995) summarize fourteen empirical studies of advisory committees that deal with environmental issues (Table 2.3). The summary includes only one case study (Hannah and Lewis, 1982) that discusses delivery of human services, dealing with a number of planning aspects. However, there is no literature available on planning and development matters within a form of planning advisory committee. Mostly, public participation literature has the broader basis of environmental aspects (James, 1999), which include planning and development matters. There are many advisory committees in Australian cities dealing with the issues of environment, conservation, parks, open-space management, and planning and development matters together.

Advisory committees differ from other public participation techniques. Typically, advisory committees consist of a small group of people whom a sponsor convenes for a specific time period. Advisory committees work as the representatives of different communities or interest groups, and envision to include their ideas and attitudes in planning proposals, issues or a set of issues (Lynn and Busenberg, 1995: 148). What distinguishes an advisory committee from other techniques, such as public hearings or surveys depends on the interaction between interested citizens and government representatives. Unlike such citizen bodies at the expert and policy levels, advisory committees are not traditionally given final decision-making powers (Lynn and Busenberg, 1995). Expert and policy-level advisory committees are more influential than other tasks-level advisory committees.

There is little difference between a task force and an advisory committee, and the two terms are often used interchangeably (Beierle, 1998). Some literature on public participation treats task forces differently from advisory committees, pointing out that an advisory committee is usually larger, longer-lived and better suited to the consideration of a broad range of issues than a task force. A task force tends to focus on one discrete issue on an *ad hoc* basis (Rosener, 1978b; Creighton, 1986, 1993). On the other hand, citizens' juries and citizens' panels are also both distinct from advisory committees.

Table 2.3 Overview of studies of advisory committees

Authors	Types of CAC	Organizations advised	Issues considered	Study methods
Priscoli (1997)	CAC involved in water resource planning	River Basin Commission	Flood plain management, irrigation and wetlands	Multiple case studies
Pierce and Doerksen (1976)	CAC created by water resource department	State department of ecology	Water resource planning	Survey-based studies, 120 people interviews from 5 CACs
Hannah and Lewis (1982)	Locally initiated CAC	City departments	City planning, and environmental concerns	Survey based studies: questionnaires, interviews, and direct observation
Stewart and Dennis (1984)	Air quality planning	Government and Governor	Clean air ACT	Individual case study and direct observation
Lynn (1987)	Hazardous waste management	City Council	Risk assessment of hazardous waste management	Multiple case studies; interviews and document analysis
Nelson (1990)	Agricultural and environmental interests	State	Soil erosion, wetlands and farmland protection	Individual case study; document analysis of meeting minutes and agreements
Ross and Associates (1991)	Hazardous Waste Advisory Council	State Government	Regional hazardous waste management system	Participate observation, and document analysis
Scrimmeour (1993)	Site specific Advisory bodies	Department of environment	Clean up of sites; hazardous materials and radioactive waste	Multiple case studies; interviews with sponsors; document analysis
Dickinson (1999)	Policy level environmental advisory bodies	Municipal council	Flora and Fauna, conservation and natural environment	Individual case study, interviews and document analysis
Tuler and Webler (1999)	Tasks-level advisory committee	Forest department	Forest resource conservation	Semi-structured interviews, document analysis and participant observations

Source: Adapted from Lynn and Busenberg (1995: 149–150)

These two types of advisory bodies, which have also been concerned with public agencies, are usually selected randomly from the population of a region, meet for short periods (usually 3–5 months), and generally limit their deliberations to one issue (Crosby et al., 1986: 173).

The public join advisory committees to perform a variety of tasks in many contexts (Webler, 1992). Some are project-specific, such as committees formed by nuclear clean-up programs (Landre and Knuth 1993a), local government for environmental and planning (Lynn and Busenberg, 1995), and water resource planning (Syme and Sadler, 1994). Therefore, advisory committees have different types of benefits and limitations in order to receive a wide range of advice in the decision-making process. Webler (1992) notes that committee members on environment issues are chosen from a diverse background of professionals and laypersons, such as farmers, doctors, land developers, environmentalists, business people, teachers, politicians and clergy. During the tenure of advisory committees, members are encouraged to return to their 'constituencies' and receive feedback – thus a sort of representativeness is established. Thus, the advisory committees perform liaison services between the project implementation authority and the community (Harding, 1998).

However, some authors are critical of the advisory committees for not being an open and fair form of public participation, as selected members of a committee may be acting with vested interests rather than in the interest of the community (Smith, 1993; Birkeland, 1999; Beder, 1999). A further criticism of advisory committees is that the proponents may form them in order to 'win' the support of neighborhood and community groups. More in the past than the present, proponents have commonly established committees as a means to 'educate' the community, labelling it 'public participation' even when there was no effective input to discussion from the community members (Beder, 1999). Also, advisory committees can become 'buffers' between the community and organizations implementing the planning decisions. Creighton (1993) and Thomas (1995) suggest several unique benefits that an advisory committee can offer to an agency. They observe that through advisory committee a consensus decision may be reached quickly rather than through public meetings, or separate negotiations with concerned groups. The advisory committee can serve as a communication link to the constituents they represented and as a means for building consensus among conflicting groups.

In evaluating advisory committees, importance should also be given to others who are involved in planning decisions and share similar information with advisory committees (Gleeson and Low, 2000). How their interactions affect the planning decisions should also be examined. How they share common information and interpret it for rational application to the planning initiatives should be considered as well. The planners, who interact with the advisory committees during the consultation process, have similar roles in the planning decision-making process. Pross (1992) has developed a policy community model for giving a broader context to the concept of 'stakeholders'. The policy community model could be translated into what Gleeson and Low (2000) termed a 'planning community' concept.

2.5 Adopting the policy community model in planning

Different individuals, agencies and interest groups who work as major actors in different fields of government activities are considered as the policy community (Pross, 1992: 119). A conceptual diagram of the policy community model is shown in Figure 2.5. Pross (1992: 119) outlined another concept namely the 'policy network'. Policy network is the relationship that forms among the actors of particular interest that is important to the policy

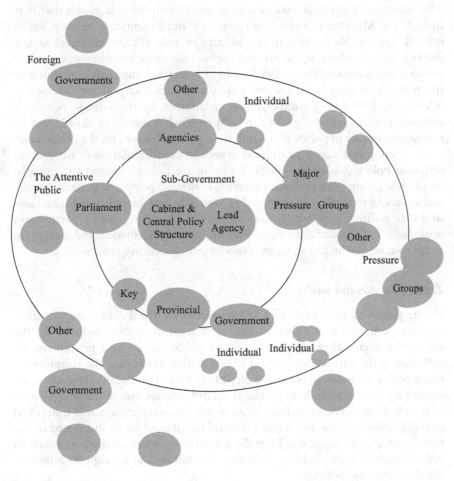

Figure 2.5 The policy community model.

Source: Pross (1992: 123)

community. He explains that policy networks are composed of like-minded people within the policy community (Pross, 1992). However, the network members may cross the boundaries of 'policy communities' and may establish communications with other interested proponents on the same issues they are dealing with. The network is composed of many government policy-implementation agencies, pressure groups, community groups, media people, individuals and academics, who have an interest in a particular policy field and attempt to influence those policies (Pross, 1992). Pross (1992) divided the people in the policy community into two groups, viz., the sub-government and the attentive public.

2.5.1 The sub-government

The sub-government consists of a very small group of individuals that may include the Minister in charge of the policy-implementation agency, senior officials responsible for this field and state or federal representatives of few interest groups whose opinions and support for implementing government policies are essential (Pross, 1992). These interest groups can contribute in the policy planning in two ways: policy advocacy and policy participation (Coleman, 1985). Policy advocacy is an attempt by the outside people of sub-government interest groups to influence what should or should not be the subject matter of policy planning. Policy participation, on the other hand, is the active participation of interest groups in the formulation or implementation of policy (Coleman, 1985). To influence policy decision successfully, an interest group must maintain internal cohesion; possess or generate information about the policy process, public opinions and the policy field; and also mobilize political support (Coleman, 1985). Therefore, policy participants must collect and coordinate a complex variety of information and activity in order to take part in the sub-government policy-making process.

2.5.2 The attentive public

Public policy is not made and implemented by the public; instead, government officials and political leaders determine public policies in the sub-government. However, the attentive public does play a pivotal role in influencing the nature and context of that policy, as they are the recipients of those policy decisions (Cole and Cole, 1983). Pross (1992: 121) claims that the attentive public may include people from different spheres of the society such as government agencies, private institutions, pressure groups and individual citizens. These people and agencies could be affected by or interested in the policies of specific agencies. They do not directly and regularly participate in policy- and decision-making processes, but follow and attempt to influence the decisions and policies.

Although the attentive public lacks the statutory power to be included in the sub-government structure, they play an important role in the policy-making process. The primary function of the attentive public is to continually review and inject differing views into policy-making. It provides the policy-makers with an element of diversity, draws attention to the inadequacies of proposed policies and often introduces new ideas based on its experience (Pross, 1992: 129). The public can contribute to policy-making by using communication methods such as consultation in terms of giving and receiving of advice and participation through policy-related workshops and documents (Pross, 1992).

2.6 The citizens advisory committees: A policy input mechanism

The CACs are statutory parts of the policy community as outlined by Pross (1992), and their primary role is to advise concerned agencies. They acquire status depending upon the scope of their mandate in the consultation process. Consequently, there are many different types of advisory committees whose terms of reference vary widely, and whose mandate can range from partial to comprehensive scale (Filyk and Cote 1992). In general, statutory advisory committees have access to all members of the policy community, including the project implementation agency, the executive, the sub-government and the wider public (Filyk and Cote, 1992). In a broader sense, there are three types of advisory committees depending upon their role in the participation process. These are *expert-level committees, policy-level committees* and *site or task-specific advisory committees* (Long and Beierle, 1999). The expert-level committees are designed to provide outside technical advice on issues relevant to the function of agencies. Policy-level committees advise more on value-laden, social dimensions of policy. For Long and Beierle (1999), generally, different stakeholders' substantive inputs are provided by the policy-level committees. Such committees 'act as a sounding board for the acceptability of policies and provide some amount of democratic legitimacy to decisions' (Long and Beierle, 1999: 6–7). Regulatory negotiations and policy dialogues are designed to generate the substance of environmental decisions through consensus among various stakeholders.

While expert and policy-level advisory committees typically deal with national issues strategically, the task-specific advisory committees deal with a defined geographic area. Many agencies have moved toward more locally based advisory committees in recent years. Trends in environmental management issues such as in ecosystem management, community-based environmental protection, integrated watershed management, as well as attention to environmental justice, have all necessitated the more active involvement of local communities and interests (Long and Beierle, 1999).

At the task-specific level, participants are likely to be 'closer to the people' than in policy-level committees (Long and Beierle, 1999). As committees become more issue-specific, boundaries between stakeholder jurisdictions are also likely to be more blurred. In some cases the leader of a community-based organization may also be a business or home-owner in the area; thus, the community leaders may expect a higher level of interest in the outcomes of the decision. Long and Beierle (1999) note that at the site-or task-specific level, it may be easier to identify poorly funded or poorly organized advisory groups. However, community consultation through the task-specific approach demonstrates a more direct model of democracy (Long and Beierle, 1999). Policy-level committees represent a more traditional pluralist approach to decision-making.

2.7 The urban planning policy community in the context of ACT

The concept of 'policy community' has been developed to describe the environmental issues on which public policy decisions are made (Filyk and Cote, 1992; Dickinson 1999). The rise of environmentalism in the past few decades has produced various advisory groups to counsel on environmental affairs (Long and Beierle, 1999).Accordingly, the 'policy community model' has been applied to evaluate municipal environmental advisory groups (Dickinson, 1999). The municipal advisory groups are mostly expert-level committees giving advice to the municipal government on environmental issues, but this model has not so far been applied in task-level planning advisory committees. In order to evaluate advisory committees in the public participation process and their roles in planning decisions, the policy community model discussed earlier must be adapted at the urban level of 'planning community'. Gleeson and Low (2000) emphasizes the importance of a systematic and organized 'planning community' for 'better cities' programs, but they have not given any theoretical foundation for analysis in planning situations. Before a description of the way the policy community is defined in the planning decision-making process, a brief overview of the structures and process of planning agencies in the ACT is required.

Planning and Land Management (PALM) was the official name of the Planning Authority of the ACT Government and can be compared with the traditional structure of city councils as part of local government. Note that name of the planning authority has changed couple of times. First it was changed to ACT Planning and Land Authority (ACTPLA) and then changed to Environment, Planning and Sustainable Development Directorate (EPSDD).

The following section describes the structure of the ACT Government and provides information on planning components. The focus is on Planning Authority, so no detail is provided about other levels of local government in Australia such as city councils and shire councils in various states and territories.

Table 2.4 Advisory committees on planning and design matters in ACT

Expert Advisory Committees	
National Capital Design Review Panel (NCDRP)	The NCDRP is a joint initiative between the ACT Government and the National Capital Authority to provide independent, expert and impartial advice for significant development proposals across the city. The NCDRP reviews a diverse range and scale of development proposals; including apartment, mixed-use and commercial development; public infrastructure projects; public parks and spaces; master plans and estate development plans
Policy-Level Advisory Committees	
Planning and Development Forum (P & D Forum). The name of the Forum has changed to 'Environment and Planning Forum'	An industry and community-based group, which takes a strategic overview and promotes input to the management of the ACT Government's planning and development policy agenda.
	The Forum advises the Minister for Planning and Planning Authority on current planning policies and initiatives, the development of any new planning policies and initiatives, communication of those policies to Canberrans, and policy evaluation, including responses to community and industry consultation processes. The Forum is made up of 11 members, who represent industry groups, professional groups, conservation councils, national trusts, ACT councils of social services and community groups.
Task-Level Advisory Committee	
Local Area Planning Advisory Committees (LAPACs). The new name is EPSDD Consultative Committee	LAPACs are committees of residents, business people and community groups who provide community advice to the Minister for Planning on planning and development matters. The main role of the LAPAC is to comment on the planning and design matters referred to committees that include publicly notifiable applications for commercial, retail, industrial, multi-unit development, preliminary environmental assessments (PEA) and other major development applications.

Source: Compiled from ACTPLA (2003d); EPSDD (2020).

2.7.1 Planning authority – Environment, planning and sustainable development directorate

EPSDD officially describes its functions, responsibilities and the area of achievement in every annual report (EPSDD, 2020). It says EPSDD manages 'a wide range of policies and programs to deliver on the ACT Government's key planning, land management and environment priorities' (EPSDD 2020: 18). EPSDD comprises various sections and branches, such as the Territory Plan, Planning and Development Policy and Land Information, Building Services Branches, Parks and Conservation and Design and Review (Figure 2.6). However, two branches, the Territory Planning Branch and the

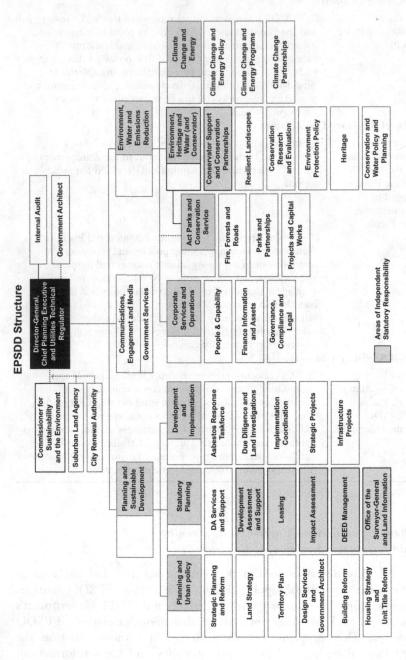

Figure 2.6 Organizational structure of planning authority.

Source: Compiled from EPSDD website; often change regularly. https://www.environment.act.gov.au/__data/assets/pdf_file/0003/581727/organisation-chart.pdf [online] 20 September 2021

Design Review Branches are mostly responsible for carrying out planning activities and consultation programs across the ACT.

The Territory Planning Branch develops long-term strategic urban and rural planning on a metropolitan and regional level. It manages and reviews the Territory Plan; prepares master plans, development control plans and guidelines for development applications; monitors urban development trends; and evaluates environmental assessments of the impacts of development proposals. The branch has responsibility for transport planning and develops sustainable development policies for the urban revitalization programs. It also investigates infrastructure requirements for the release of land, and prepares and manages capital works programs (EPSDD, 2020). The main responsibility of this branch is to examine continuing proposed variations to the Territory Plan. These proposals come to this branch after a rigorous consultation process with various planning stakeholders and advisory committee

The Development and Management Branch manages the lodgments, assessment and determination of development proposals, prepares new leases, and manages deeds of agreement. It administers the *Unit Titles Act 1970,* enforces lease provisions through compliance action, provides support for advisory committees and reviews and prepares legislation associated with its development management role. This branch normally conducts public consultations programs with various statutory and non-statutory planning stakeholders.

2.7.2 Role of interest groups

Apart from the planning stakeholders participating in various planning committee meetings, the various interest groups on planning also attend in meetings. The presence of interest groups at meetings is not regular. They attend the meetings in greater numbers when they feel the planning proposals are important and will affect them. However, some interest groups regularly attend meetings. They are non-professional interest groups concerned with environment, planning and development matters in their areas or elsewhere in Canberra, such as Planning ACT Together (PACTT), Save the Ridge, O'Connor Ridge Parkcare, Friends of Black Mountain, Aranda Bush, Belconnen Community Council, Old Narrabundah Community Action Group, North Canberra Community Council, and many other resident associations. These community-based groups can be characterized as *collective rights groups* (Beierle, 1998) since they are the bodies comprising various voluntary community action groups and resident associations. Considering the nature and activities of these groups, Lightbody (1995) has described them as noisy, negative, and reactionary. These groups are noisy because they exploit the media to gain attention to counteract institutionalized power and resources, negative because they always pose negative attitudes towards any development proposal if they think the proposal is detrimental to a part of the city, and reactionary because they are not built into the policy process (Lightbody, 1995).

Some professional groups also participate in the Committee meetings to express their concerns on planning and development matters. These groups have no legal status to be consulted regularly, but PALM often consults them at formal meetings. Sometimes, the Minister for Planning also comes to talk with LAPAC members in the meetings. The professional groups are mostly architecture firms, developers' associations and real estate agents, and can be characterized as *institutionalized interest groups* on planning and urban revitalization programs (Lightbody, 1995). They have no voting power in the final decision-making in the Committees, but they are given ample opportunity to participate in the discussion and submission of their concerns to planning authority. Lightbody (1995) notes that these interest groups use political resources such as social standing, professional knowledge and corporate wealth in a continuing policy-dialogue with city councils and senior planners. Pross (1992) observes that those institutionalized interest groups are located within the sub-government structure of an urban policy community.

The above sections provide an overview and description of the principal members of the 'urban policy community' in Canberra involved in planning and development matters. Although the structure of a policy community varies from city to city and from policy field to policy field, some generalizations can be made about the urban planning policy community and people influencing planning decisions. Figure 2.7 provides a visual representation of the conceptual urban planning policy community in Canberra. The boundaries between the sub-government and the wider public are not rigid. Both groups may be occasionally involved depending upon the intensity of the planning and development matters. Pross (1992) describes the sub-government as policymaking body that consists of government agencies, and the attentive public are those who draw attention to inadequacies in government policies. However, this book adapted the 'attentive public' into 'wider public' in the context of consultation process of advisory groups where enthusiastic individuals come alone to comments on issue-specific planning initiatives only.

The sub-government is made up of Planning Authority's standing committee on planning and environment, the Territory Planning Branch, and the Development and Management

Branch, with responsibility for developing Master Plans, section Master Plans, multi-unit development proposals, redevelopment of goup centers, variations to the Territory Plan and carrying out consultations. The interest groups are development proponents, political parties, and community-based organizations interested in planning and development matters. The wider public includes individuals interested in planning matters. Overlapping the boundary between the wider public and the sub-government is the advisory committees (LAPACs), a means by which members of the public can seek to participate in and influence planning decision-making. Figure 2.7 shows

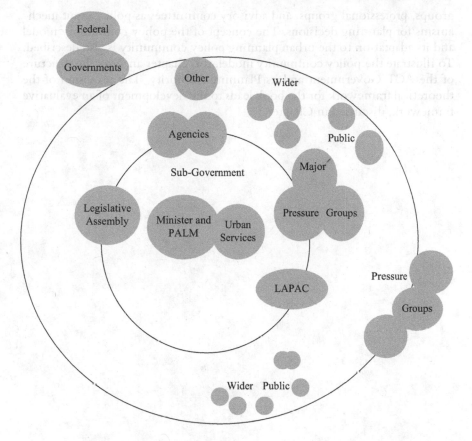

Figure 2.7 Urban planning policy community.

Source: Adapted from Pross (1992)

Planning Authority's executive, department of urban services, standing committee on planning and environment, the Territory Planning Branch, Development Management Branch, interest groups, and LAPAC itself. The discussion of Planning Advisory Committee case studies in Chapters Five and Six, while not strictly limited to these key actors, does focus primarily on the roles on participants in the consultation process and their perceived understanding of the fairness and effectiveness of public participation in planning decisions.

2.8 Chapter summary

This chapter discusses existing planning theories in relation to public participation. It also presents the models, approaches, and techniques of public participation in planning. It describes the structure of policy community models including structure of sub-government, the wider public, collective interest groups, professional groups, and advisory committees as policy input mechanisms for planning decisions. The concept of the policy community model and its adaptation to the urban planning policy community is also described. To illustrate the policy community model, this chapter analyses the structure of the ACT Government and its Planning Authority. The provision of the theoretical framework for the book leads to the development of an evaluative framework, discussed in Chapter 3.

3 Theoretical framework for evaluating public participation in urban planning

3.1 Introduction

Many planning agencies have experience in conducting public participation programs and many methods of participation have been suggested and implemented. However, important questions as to the most effective methods of public participation remain and there seems to be a general need for more systematic knowledge about ways to make public participation successful. Despite the widespread interest over the last several decades, no consistent method has so far emerged for evaluating the factors of success and failure of participatory processes concerning planning and development issues (Davies, 1998). Searching the literature on public participation in planning showed that there was little comprehensive evaluation. Rosener (1981: 583) noted that the concept of public participation is complex and value-oriented, thus there are no agreed-upon evaluation methods and no widely held criteria for assessing the success or failure of a participation process.

This chapter discusses the existing evaluation models, which provides an insight into specific criteria that may lead to developing an acceptable process for other approaches to evaluate participation.

3.2 Approaches to the public participation evaluation

A number of authors have recognized the need for a systematic evaluation of participation programs (Homenuck, 1977; Rosener, 1978a, 1978b, 1981; Sewell and Phillips, 1979). Homenuck (1977) notes the importance of systematic evaluation, which will provide a learning framework to improve the process and avoid past mistakes. Considering the barriers to public participation evaluation and the problems of identifying issues of evaluation, the existing literature describes three major approaches: user-based evaluations; theory-based evaluations and process-based evaluations (Palerm, 2000; Raimond, 2001).

3.2.1 User-based evaluations

The principle of *user-based evaluation* is that different participants will have different goals. Instead of trying to reconcile these goals, researchers have

DOI: 10.4324/9781003122111-3

developed evaluations based on a questionnaire that includes the conflicting goals of citizens and agency staff (Chess, 2000: 774). The user-oriented evaluation provided information on whether the participation goals and objectives of all participants were achieved in the workshop process. The information was also used to develop an overall effectiveness measure for evaluating the workshop. This measure focused on goals that were shared between participating groups; if all the shared goals were met, the participation activity was evaluated as effective. According to Rosener (1978b: 585), the evaluation resulted in information that was comparable and that allowed for generalizations about the workshop process.

3.2.2 Theory-based evaluations

Theory-based evaluations rely on criteria that are based on theories and models to evaluate public participation efforts (Raimond, 2001: 27). The first example of theory-based evaluation is using a 'social goals' framework (Beierle, 1999). The second example of a theory-based evaluation, based on 'fairness and competence' (Webler, 1995) is presented here because it provides a good example of the work of researchers in the United States, Germany and Switzerland, and offers insight into the results of experiments with novel approaches to public participation (Raimond, 2001: 29). The third example described here is based on the 'Theory of Social Psychology of Procedural Fairness' (Lauber and Knuth, 1999), and describes the operational process of consultation in natural resource management, mostly in the form of Citizen Advisory Committees (Landre and Knuth, 1993b; Lauber and Knuth, 1998, 1999).

The social goals framework

The social goals framework focuses on the evaluation of participation mechanisms intentionally instituted by government to involve lay people, or their representatives, in decision-making issues. Raimond (2001) notes that this evaluative framework determines whether participatory programs are working or not, how they can be improved, which methodologies work best for particular needs and ultimately whether participatory programs justify the commitment of public and private resources.

Social goals, according to Beierle (1999), are those goals, which can be applied to many participatory mechanisms. He identified five evaluation goals: educating and informing the public; incorporating public values into decision-making; improving the substantive quality of decisions; increasing trust in institutions and reducing conflict.

Social psychological theories of procedural fairness

Another theory-based evaluation of public participation is based on social psychological theories that have made major contributions to our understanding of how participants judge the fairness of the consultation process. (Lauber and Knuth, 1999; Laird, 1993; Thibaut and Walker, 1975; Tyler, 1989; Tyler and Griffin, 1991)

Lind and Tyler (1988) have characterized the criteria of fairness into two sections, distributive and procedural fairness. 'Distributive fairness refers to the fairness of decisions, and procedural fairness refers to the fairness of the processes used to produce these decisions' (Lauber and Knuth, 1999: 20). The criteria used for evaluating procedural fairness have received considerable attention due to their relation with participants' satisfaction with the existing consultation procedures.

Using meta-criteria: Fairness and competence

Another theory-based approach to evaluation was proposed by Webler (1995). Using the Jürgen Habermas's (1984) concepts of *ideal speech situation* and *communicative competence* (Habermas, 1987), Webler developed evaluative criteria of environmental decision-making, based on two meta-criteria: *fairness,* and *competence.*

Webler (1995: 79) defines *fairness* as distribution of sufficient opportunities for the participants to act meaningfully in the participation process. When people come together with the intention of reaching, understanding and making public decisions in a fair process. To him, four fundamental actions for the participant must be available. These are to *attend* (be a participant in consultation meeting); *initiate* discussion (make statements on issues); *debate* in discussion (ask for clarification, challenge, answer and argue) and *decide* in the decision-making (resolve disagreements). These are the four *needs* of a fair process and are relevant to each of the three *activities* that constitute a public participation process: agenda and rule making, moderation and rule enforcement and discussion.

Webler (1995) has discussed the above criteria for evaluating the fairness of public participation process. Fair participation in agenda-setting and rule-making means that all participants in the consultation process should have the same opportunity to take part in activities that determine the agenda and rules for discussion. Fairness in the moderator and rule enforcement means that everyone in the process should have an equal chance to suggest moderators and a method of facilitation. Fairness in discussion and debate is making sure that everyone has an equal chance to discuss an agenda and to influence the final decisions (Webler and Tuler, 2000).

Webler (1995: 81) defined *competence* as the performance of individuals in the consultation process, compared to what can be reasonably expected of them, given the current information and knowledge available.

Webler (1995) speaks of competence as being capable of contributing to the definition of the collective will through participation. Therefore, participants are not only competent in understanding terms, concepts and definitions but also able to share the social construction of reality. Competence thus relates to 'psychological heuristics, listening and communication skills, self-reflection and consensus building necessary for understanding and agreement to emerge from the participation process' (Webler, 1995: 58).

3.2.3 Process-based evaluations

A *process-based evaluation* depends on perspectives and criteria set aside by the researcher, which are believed to be pertinent for specific conditions of issues that are being evaluated (Beierle, 1999). The most common criteria evaluated in the process-based approach were the perceptions and attitudes of participants toward the existing context and process of public participation (Dickinson, 1999). This approach broadly identifies criteria that include the participant's perception of the decision-making process; the agency's perception of the consultation process; assessing whether participation resulted in a change in policy; comparing the objectives of the participation process with the outcome; process analysis to ascertain how participation was implemented and functioned; perceptions of the general public; comparing overall agency goals and objectives with outcomes of the participation process; cost efficiency; relationships among stakeholders and feedback analysis (Cole, 1974; Pierce and Doerkson 1976; Homenuck, 1977; Ertel, 1979a, 1979b; Cole and Caputo, 1984; Hutcheson, 1984; Hutcheson and Prather, 1988).

Examining of the literature on planning makes it clear that neither process-based evaluation nor theory-based evaluation alone can be recommended as ideal guidelines for the evaluation of public participation in planning. Considering the current limitations, this study proposes to integrate process and theory-based criteria for evaluating public participation in planning. The following section is a theoretical framework developed for evaluating public participation in planning. Most evaluation methods followed a method of public participation process, thus this study used planning advisory committees for the evaluation.

3.3 A theoretical framework for evaluating public participation in planning: Fairness

This book has developed a theoretical framework for evaluating public participation in planning integrating process and theory-based evaluation. Combining the principles of social goals (Beierle, 1998) and issues of the context-process-outcome approach (Dickinson, 1999; Smith, 1993) with those of fairness and competence (Webler, 1995; Lauber and Knuth, 1999) produces ten evaluative criteria, twenty-six sub-criteria, and sixty-four questions on whether the process is fair and effective. Figure 3.1 shows the essential criteria for evaluating the fairness and effectiveness of public participation in planning (detailed questions for all criteria are given in Table 8.1 and Table 8.2 in Chapter 8). The framework integrates the available theories and discusses pre-determined criteria (i.e., what criteria should evaluate) and

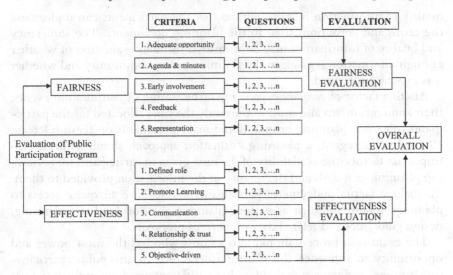

Figure 3.1 A theoretical framework for evaluating public participation in planning.

existing processes (i.e., what criteria currently exist to evaluate) in order to identify the strengths and weaknesses of public participation.

The aim of the framework is to recommend evaluative criteria, which are likely to work better. The proposed criteria can be applied to many public participation mechanisms in general and CACs in particular. However, issues and places may vary in context and therefore need consideration for amalgamation with existing criteria. Evaluation criteria should be framed in a way that enables evaluators to consider local issues and to fit them in the evaluation process.

3.3.1 *Adequate opportunity*

One of the important criteria for evaluating fairness in public participation is that of *adequate opportunity* (Octeau, 1999). Adequate opportunity is an assessment technique from the point of view of an individual or group being given enough opportunities for each person to protect and express legitimate interests and contribute to the development of consensus for a planning decision (Renn et al., 1993). Individuals in a group must be able to initiate actions that enable them to accomplish their legitimate goals, whatever the nature of their representation in the participation process be (Webler and Renn, 1995). In a fair participation process, participants must feel empowered and valued for making recommendations and evaluating the rationale of others assertions on planning and development matters (Webler, 1995). Octeau (1999) notes that participants feel empowered once they are given sufficient and good-quality information. Sufficient information is needed to understand all factors in the continuing issues of planning (Webler et al., 2001); and the quality of information is important so that the participants can understand the terms and definitions used in the planning documents. The sufficiency and quality of information may be measured through evaluation of whether enough information is given to the planning policy community and whether it is easy to understand.

Another factor of *adequate opportunity* is the timing of consultation events, their announcements and most importantly the time allocated for the participants to review planning proposals and make comments on them (Octeau, 1999). In this regard, a planning evaluation approach should address two important factors: the availability of the time given to participants for reviewing planning proposals and the quality of the information provided to them. To make a factual judgement, participants should have adequate access to planning documents such as various planning Acts on land use planning, design guidelines and related data on current proposals.

The evaluation process should also address who has the most power and opportunity to influence the planning decisions. In any public participation process, different stakeholders have different levels of influence over decision-making process (Abbott, 1996; Octeau, 1999; Dickinson, 1999; Webler and Tuler, 2000), although some believe that equalizing the amount of power to influence the final outcomes is the major criterion for evaluating the fairness of public participation (Abbott, 1996; Octeau, 1999). To evaluate this criterion, all discourse participants should be asked to evaluate their understanding of influence in the final decision-making and to identify the position of all the planning policy community in the decision-making process. This will give an indication of who has the greater influence on the planning decisions. Three sub-criteria have been identified for evaluating the criterion of 'adequate opportunity': there should be equal opportunity to participate meaningfully, everyone should have equal access to information and all should have an equal opportunity to influence the final decision.

3.3.2 Representation

It is important to draw a distinction between *representing interests* and *being representative of interests*, when evaluating a participation process (Petts, 2001). For representation on a planning committee, the participation process should make considerable efforts to select individuals who may be representative of a wide range of interest in the community, rather than people simply representing some self-selected subset (Tuler and Webler, 1999). Particular caution should be exercised with regard to marginalized segments of the community or recruitment of intelligent and motivated people with legitimate interests in planning and development (Rowe and Frewer, 2000), so recruitment of participants should focus on personal activities that might indicate an interest. To ensure a fair representation, possibly affected communities and individuals should be included in the consultation process, but consideration should be given to recruiting non-professional planning experts who are not planners by profession, but are involved with planning and design matters or have shown interest in the planning process. These should be either selected or appointed using the planning agency's discretionary power.

It is assumed that commenting on development applications requires understanding of planning laws and Acts, most importantly understanding of development approvals, without a working knowledge of which a participant cannot make informed comments on development applications. Similarly, proponent of development should also be represented in the planning meetings to give their views. Rowe and Frewer (2000) note that one approach to achieving good representation for a regular advisory committee is to select a stratified sample from the affected community. They claim that the appearance of any bias in sampling may undermine the credibility of the committee (Rowe and Frewer, 2000). Only one sub-criterion has been proposed to evaluate the criterion of representation that includes all the planning policy community: the approach should ensure greater representation from all the planning policy community.

3.3.3 *Agenda setting and minutes taking*

Agenda setting is much the same as defining the problem (Webler, 1995). Webler (1995) broadly generalized this criterion as internal fairness based on some process criteria that are usually followed in the discussion, such as roles in agenda setting, determination of moderators and agreement on discussion, operational procedures, and final decision-making procedures within the planning committee. However, this does not mean that everyone should necessarily play an active role in making the agenda. Webler (1995) has noted that an agenda can unfairly influence the ensuing discourse by not allocating ample time for discussion, or by framing a topic and putting an issue in such a way that most of the members will have no chance to contribute effectively on the particular development application. Other important issues, which Webler (1995) has not mentioned, have been raised regarding this criterion, which involve ensuring that all planning matters are included on the agenda. Four sub-criteria have been proposed to evaluate this criterion: there should be provision for everyone to have the opportunity to define planning and development issues before formal development applications are lodged, all planning and development matters should be included in the meeting agenda, the meeting minutes should include comments from all planning stakeholders, and the minutes should be available for public perusal and should be distributed among planning stakeholders in good time.

3.3.4 *Early involvement*

One aspect frequently discussed in the public participation literature is the stage at which the public should become involved in policy matters (Rowe and Frewer, 2000). The consensus seems to be that public participation should occur at the beginning rather than at the end (Middendorf and Busch, 1997). There should be a formal structure in which public and community groups should be involved to express their concerns on development applications. The general thumb rule is that the public participation should occur early in the planning process to make it meaningful (Brody et al., 2003: 250). Early involvement allows plan to reflect community views and preferences.

Two sub-criteria have been proposed to evaluate the criterion of early involvement: everyone should have an opportunity to be involved during the conceptualization of the planning proposal, and everyone should have an opportunity to be involved in inter-departmental planning meetings.

3.3.5 Feedback

Feedback is an important criterion for evaluating the fairness of a public participation process. This criterion examines the responsiveness to communication among all planning stakeholders, including planning agencies, development proponents and other participants, in order to respond to each other's needs in such a way that all are satisfied with the actions. People participating in the consultation normally expect feedback from the implementing authority showing that they are being heard, and their comments and advice are given adequate attention. The planning authority should also give the participants convincing reasons for rejecting or modifying their comments and advice. Two sub-criteria have been proposed to evaluate the criterion of feedback: feedback should be provided to the participants in the consultation process, and explanations should be given to participants for the acceptance, rejection and modification of their recommendations.

3.4 Theoretical framework for evaluating public participation in planning: Effectiveness

3.4.1 Defined role

It is essential to ensure that there is as little confusion and dispute as possible regarding the scope and role of the stakeholders in the participation exercise (Landre and Knuth, 1993a). The roles of the various stakeholders can change with different approaches to public participation (Octeau, 1999). Particularly in the case of stakeholders' committees, roles can vary and therefore must be clarified. For example, planning advisory committees consist of many people involved in planning policy, ranging from development proponents to residents and community groups. The use of experts for technical clarification and to explain the government position on planning proposals can support this type of planning committee. The development proponents can also play a similar role. Sometimes, the planning agencies may have discretionary power to nominate experts or enthusiastic individuals to the committee. The roles of such people who are implicitly or explicitly involved in the planning process should be clearly defined before the consultation activities begin (Cormick et al., 1996).

The criterion of 'defined role' should illustrate the possible roles that may be played by the planning policy community in decision-making. All those involved in planning policy should also accept their defined roles in the consultation process. There should be a clear description of expected outcomes of the participation as well as of the way the procedures work (Rowe and Frewer, 2000). All participants in the consultation process should have adequate understanding of their roles, Terms of Reference, and related planning laws and Acts. When members start their tenure, the planning agency should provide all members a guide with adequate description of their roles and clear statement of the Terms of Reference, because without a clear statement of their roles members may feel disempowered and disenfranchised (Octeau, 1999). The evaluation of this criterion should ascertain whether planning stakeholders are effectively playing their defined roles and participants should give their perceptions on whether the defined roles would lead to an effective consultation or require any changes.

Three sub-criteria have been proposed to evaluate the criterion of defined role: well-defined Terms of Reference should be provided for the consultation process, Terms of Reference should define all stakeholders' roles in the planning decision and all stakeholders should understand their roles clearly and succinctly.

3.4.2 Promote learning

Public participation processes are learning opportunities for all participants (Octeau, 1999).They should be treated as a platform of interpersonal communication for sharing information, but not as a battleground for conflict of interests (Landre and Knuth 1993a). An effective participation process should therefore promote education and learning activities (Beierle 1998, 1999; Octeau, 1999). Once the participants are well-informed about process roles and the relevant planning laws and Acts, they can carry out the roles envisaged in major planning legislation for identifying violations of planning laws, and whether development applications comply with the required design guidelines (Beierle, 1999). In the context of the stakeholders' committee, the planning agency should organize planning workshops and orientation programs to introduce salient features of the existing planning law, design guidelines and the procedures for development approval. The introduction of orientation programs will help the participants to understand planning legislations and identify any violations of planning law by development proponents, so they will gain sufficient knowledge of planning and development approval. This will enable them to deliberate issues and formulate alternatives to the development proposals.

To achieve this goal, and to enable participants to understand planning laws and the process of development approval, the planning authority should use accessible language in planning documents. To evaluate this criterion, it is important to know how many participants were actively involved in promoting learning or took advantage of the information.

Three sub-criteria have been proposed to evaluate the criterion of promote learning: there should be effective outreach programs for the participants, introducing planning law and design guidelines that are essential for compliance with planning proposals, and there should be a clear plan that includes ways in which the participants can learn about the process.

3.4.3 Communication

Communication is an integral part of the community development process (Ashford, 1984). There are some mechanisms normally adopted by the planning authority to communicate with rest of the policy community. Accordingly, government agencies have used a variety of methods of communicating in sharing relevant planning and design information with the stakeholders (Dandekar, 1982). But the questions remain as how effective have they been and does the participant effectively respond to the planning information provided in development applications? Communication of the planning information becomes an important criterion for making informed comments on planning proposals. Citizens who are involved in the planning process often face difficulties in understanding spatial data, which may be in the form of digital or paper products displayed on maps and plans (Obermeyer, 1998). Misapprehension of the spatial data and complications in understanding can lead to mistrust amongst planning stakeholders (Craig and Elwood, 1998). Without effective communication and understanding of planning information, the participants cannot make informed comments. Geographic Information Systems (GIS) have been used for the public participation process to establish effective communication with spatial information among the planning policy community (Craig et al., 1999).

It is apparent that planning participants with planning and development matters, and regularly discuss development applications. The development applications are accompanied by maps and plans, so development proponents bring maps and drawings to almost every consultation meeting. But are the participants able to respond effectively to displayed maps and plans? Do they understand what the developer presents? Do participants have adequate understanding of plans? How is geographic information brought to participants for making comments on it? How are maps, plans, tables and spatial information given to the participants? The planning agencies should make use of Internet facilities for communicating with their planning stakeholders. The Internet and other forms of electronic communication become effective modes of providing relevant information to the stakeholders, thus creating an effective communication (Kingston et al., 2000).

Three sub-criteria have been proposed to evaluate the criterion of communication: a process should be provided for communication among all planning stakeholders, there should be a formal structure for communication between the planning agency and planning stakeholders and there should be appropriate ways to use and display geographic information for effective communication.

3.4.4 Relationship and trust

In most of the literature the criteria of relationship, trust and conflict of interests are discussed separately. However, these three criteria could be merged for analyzing a consultation process (Raimond, 2001). In analyzing the positive aspect of these three criteria, it would be seen that good relationships among the stakeholders would reduce conflict. On the other hand, if the relationships are adversarial and hostile, the consultation will reduce trust among the participants, leading to conflict over the issues (Raimond, 2001). In almost all conversations between government staff and citizens, lack of trust or a perceived lack of trust in government agencies is raised as a cause for conflict (Long and Beierle, 1999). The absence of trust can cause the committee members to question many aspects of government decision-making, including the result of the planning and design assessment of a particular development application. Distrust among the members generates adversarial relationship with each other that often causes conflict among them on planning decisions.

All planning stakeholders should have amicable relationships with each other, respect others' views, and be prepared to accept the reasons for other people's assertions. They should also have to resolve conflict over decisions in an amicable way. Discourse participants must include people from diverse backgrounds, such as resident's community leaders, business people and agency representatives. All are engaged in making comments on a particular planning topic, and should have established a trustworthy relationship and conflict-free environment.

Three sub-criteria have been proposed to evaluate the criterion of relationship: there should be a responsive relationship among planning stakeholders, there should be a congenial atmosphere free from personal attacks and trust should be established towards the chairperson conducting consultation meetings.

3.4.5 Objective-driven

The introduction of public consultation processes for planning purposes is driven by the objectives of authorities (Octeau, 1999); however, a planning authority may implement only the objectives and political promises of a government. In this situation, the planning authority may be seen as an implementing authority for those promises, not an independent authority to recommend planning approaches based on partnership between the community and agencies for the development and redevelopment of urban-areas.

The planning and management in urban areas are ultimately a question of values, as planning approaches are based on values and priorities perceived by the government and their political promises (Troy, 1999). Over time, values and priorities may be changed, if the government is changed. However, the general objective of introducing public consultation in planning and development matters is to give the community a voice in the planning decision-making process and to ensure that planning and decision-making are subject to a process of public participation. Troy (1999) notes that a government may have a planning approach, which might not be seen as appropriate by the succeeding government.

Similarly, planning participants may have different objectives and expectations from the consultation process, while success in consultation process can be measured by the perceptions of the individuals involved in the process. Accordingly, all planning policy community may have expectations and be required to evaluate their degree of satisfaction about achieving these expectations from the planning consultation process. Effectiveness evaluations are typically correlated closely with participants' satisfaction with the procedure and its outcomes (Lauber and Knuth, 1998). If respondents perceive the decision-making procedures as fair, they are more likely to perceive the decisions as fair (Lind and Tyler, 1988). Perceptions of procedural fairness are often correlated with satisfaction or support for the decision makers or authority responsible for a decision (Lauber and Knuth, 1999).

Two sub-criteria have been proposed to evaluate the criterion of 'objective-driven': the agency's objectives should be given to participants in the consultation process, and there should be provision to regularly evaluate 'satisfaction with outcome' and 'satisfaction with process'.

3.5 Chapter summary

The evaluators of public participation processes have learned through experience to develop research designs that take into account the strengths and limitations of various evaluation approaches. This book suggests that planning agencies should not be restricted to existing methods of consultation, but allow methodological pluralism in evaluating their processes. Discussion of various evaluation approaches, particularly theory-based and process-based approaches, indicates that evaluators need not select either theory-based or process-based evaluation criteria articulated by the stakeholders, sponsors and planning agency personnel. Rather, evaluators might focus on the development of methodological diversity and evaluation methods that can be adapted to the specific participation efforts. Planning advisory committees involve a variety of meetings that discuss development applications, communications among all the planning policy community and provide relevant information to the committee members for informed discussion; so a way is needed to gather feedback about planning meetings so that planning decisions can incorporate the input emerging from the discussion. However, the experience of the evaluators of the consultation process suggests that evaluations do not necessarily lead to a structured solution; instead, some recommendations may gradually change the processes practiced. The proposed framework for evaluating public participation in planning may improve current practices more strategically and coherently. With this purpose in mind, the Planning Advisory Committees were selected as a case study.

4 Methods for evaluating public participation in urban planning

4.1 Introduction

The nature of a research problem indicates the type of method that needs to be used (Field and Morse, 1991; Rahnema, 1992). This book attempted to explore individual perceptions of a particular process in order to evaluate its context and operational process. In such instances a case study approach is particularly appropriate. Yin (2017) notes that case study research focuses on unique and contemporary events but does not attempt to generalize its observations to universal truths. Instead, information about a particular process is used to modify larger theoretical questions, or to refine or readjust existing processes. This book focuses on the Planning Advisory Committee context and process to investigate theoretical questions about the ways the consultation process is carried out, and to question its fairness and effectiveness in planning decisions.

The analysis of the methodological literature suggests that qualitative research enables researchers to observe and uncover the meaning of individual perceptions on the issues with which people are involved. The focus of qualitative research is on the participants' perceptions and experiences, and the way they make sense of their lives (Creswell, 1994). The attempt therefore is to understand not one, but multiple realities of a contemporary process (Lincoln and Guba, 1985). Considering the need for qualitative data, this book uses a multi-method approach similar to that of Kaufman (1999) and Elwood (2000). The method is mainly based on the case study approach. However, the rationale of using the multi-method approach is that it generates multiple chains of evidence and the possibility of 'triangulation' of data (Yin, 2017), which is generated by document analysis, participant observation and in-depth interviewing (Elwood and Leitner, 1998; Elwood, 2000).

This chapter begins with an explanation for selecting the qualitative methodology that guided the design of the research techniques, and a description of case study selection. This is followed by a review of data collection techniques and description of interview design. Finally, the process of data analysis is elaborated.

DOI: 10.4324/9781003122111-4

4.2 Selection of methods

Many techniques are available for investigation of public participation, and each contributes distinct benefits in understanding research problems and objectives (Sanoff, 2000), which require several different data sources to analyze and structure them (Creswell, 1998). For example, consulting the residents and community about whether they feel their advice has been given adequate consideration in planning decisions, requires qualitative data from various sources, such as interviews with the participants in the consultation process, qualitative and quantitative documentary data, observation in the planning meetings, and the available literature on resident and community views of public participation (Creswell and Creswell, 2017).

Multiple methods can be useful for gathering data from various sources (Elwood, 2000); Kaufman (1999) notes that the multi-method approach has numerous benefits. Many authors advocate for using multi-method techniques to obtain a holistic picture of the topic being studied. This approach enables the researcher to link past and present situations, and explore individual perceptions to compare them with existing situations (Berg, 1989; Reinharz, 1992). For instance, one of the most difficult tasks of this research was to get information on Planning Advisory Committee's past history. No written documents are available on the history of Advisory Committees that was introduced in Canberra in 1995, except for a few comments by the Members of the Legislative Assembly (MLA). The comments are recorded in the resolutions of the Assembly, but the written comments of MLA would not adequately reflect the underlying causes for introducing a means of planning input in planning and development matters. Thus, the author had to find participants and planners involved in the days when Committee was first introduced in the ACT.

4.2.1 Qualitative approach

The selection of a suitable approach for this book began with a survey of the literature, followed by deciding on whether to employ qualitative or quantitative methods. The main source of data for this book is qualitative, since qualitative data collection techniques such as interviewing, participant observation and document analysis were deemed to be appropriate techniques for the book. As explained in the Chapter 1, analyzing individual perceptions of the participation process is complex. It requires qualitative precision to realize the participants' perceived understanding on consultation process and its influence to planning decisions. In this regard, Patton (2014) notes that the qualitative strategy used in the participation process is largely determined by the purpose of the study, the nature of research objectives, and the skills and resources available to the investigator. He also argues that qualitative inquiry is highly appropriate in studying *process,* because depicting process requires detailed descriptions of individuals involved in the processes over a period of time (Patton, 2014).

Participation through a planning advisory committee is an individual experience for participants based on a group setting. To explore and capture such individual perceptions on a particular consultation process, this book followed the qualitative approach for data gathering and data analysis. The method chosen for this study involved making choices between five qualitative approaches: phenomenological, ethnographic, constructivist, grounded theory, and case study. A phenomenological study describes the meaning of the lived experience for several individuals about a concept or phenomenon (Creswell, 1998). The intent of ethnographic research is to obtain a holistic picture of the subject of study with in-depth interviewing and participant observation (Burawoy, 1991). Constructivist research relates to constructing a theory from existing field situations (Lewins, 1993). Grounded theory is similar to constructivist theory, requiring field data for identification of the factors to be analyzed (Glaser and Strauss, 1967). The case study inquiry is based on a specific situation and analyzes the individual perceptions of participants who are involved in the process, and qualified to make comments on existing processes and outcome (Yin, 2017). This book is based on case study inquiry.

4.2.2 Case study inquiry

Case study inquiry is a distinctive form of empirical inquiry (Yin, 2017). Thus, a case may be a person, a group, a program, an event, an activity, a process, a community, a society or any other unit of social life (Creswell, 1998; Yin, 2017). Reinharz (1992: 167) argues that case study research is useful for furnishing an idea and illustrating the consultation process over time. For him, case study research also exposes the limits of generalizations by unearthing uncharted topics, first by digging into a limited case, followed by initiating proactive questions (Reinharz, 1992: 167). Yin (2017) defines case study research as an empirical mode of inquiry that investigates a contemporary phenomenon within its real-life context, especially when the boundaries between phenomenon and context are not clearly evident. The case study copes with a technically distinctive situation in which there are many variables of interest than data points. Consequently, the case study research relies on multiple sources of evidence with data needing to converge in a triangulating fashion and benefits from the prior development of theoretical propositions to guide data collection analysis.

In conducting case study research, the researcher must consider the type of case study that is most promising and useful. This means that the researcher needs to make a firm decision in advance of going to the field and collecting data about the number and nature of cases that are expected to address the research questions. For instance, the researcher must decide whether to use a single case or multiple cases prior to collecting data. (Yin, 2017). The strongest case study design involves the comparative study of two or more communities. Hakim (1987) noted that comparative case studies are a well-established design for research on local governments and policy process. However, every study is a case study because it is an analysis of social phenomena or existing theories, specific to time and place (Yin, 2017). Thus, there are several reasons for selecting a single case. First, 'single case can be based on experiment, and *many of the same conditions* that justify a single experiment also justify a single case study' (Yin, 2017). On the other hand, a single case study design can also be applied to test a well-established theory and its current applicability. The theory may have a clear set of propositions, which are believed to be true. To understand, challenge or extend the theory, propositions may exist that meet all preconditions for testing the existing theory. As such, researchers use single case study for verifying if the existing theories are correct.

A second reason for selecting a single case study is one in which the case is '*extreme*' or '*unique*' (Yin, 1994: 39), while the third reason is the one where the case is '*revelatory*' (Yin, 1994: 40). This type of case study research depends on accessibility to the case where previous accessibility was restricted.

The above are the three major reasons for conducting a single case study inquiry. However, there are other situations in which single case study may be carried out, such as conducting of a pilot case that involves more than one unit of analysis of the same case. Yin (1994: 41) terms this as 'embedded case studies', which occur when, within a single case study, attention is also given to similar subunits of a case. However, the objection to single case study design is that a case may later turn out to differ from what it was thought to be at the outset (Yin, 1994). Yin (1994: 41) further argues that the researchers need to investigate cautiously and prudently to gather maximum case study evidence by lessening the potential chances of misrepresentation. As such, single case study research should not commence unless those major concerns have been clearly covered (Yin, 1994).

This study is in the nature of a single case study. All Planning Committees were run with a single set of Terms of Reference, and a single planning agency administers all consultation processes. Thus, Planning Committees could be a single case study; however, it had six Committees in various places with different names, which performed the same roles on planning and development matters. These Committees were unique cases compared to other advisory committees, in New South Wales (NSW) and elsewhere in Australia. In NSW and its various City Councils, the Advisory Committees are formed with elected council members, along with others, whom the City Mayor feels should be on the committees.

Selection of Planning Committees as case study inquiry

Choosing a case for study is probably one of the hardest parts of conducting any research – either qualitative or quantitative (Tutty et al., 1996). Some pertinent issues are required for the selection of cases regardless of whether the study is single case or multiple cases. An ideal case is one in which entry is possible, there is a mix of process, people, programs, interactions and structures of interest; the researcher is likely to be able to build trusting relations with participants of the study; and the data quality and credibility of the study are reasonably assured (Marshall and Rossman, 1995).

The selection of the Planning Committee is based on these guidelines. The first author attended several committee meetings accompanied by students. Besides, the author have considerable experience in assisting students on topics related to public participation in urban planning, design and development.

There were six Planning Advisory Committees in Canberra (Figure 4.1): Burley Griffin, Manuka, Ginninderra, West Belconnen, Inner North and Majura. A Planning Committee is a committee of residents, businesspeople and representatives from community groups who advise the Minister for

Planning on planning and development matters. Committee was introduced in September 1995, in the hope of involving the community in consultation about the matters affecting them directly or indirectly (PALM, 2000b). The structure of the Planning Advisory committee is as follows:

1   two people per suburb: one elected, the other appointed by a residents' group. If one exists in Belconnen Committee (called Belconnen LAPAC), one person per suburb is elected or one residents' group nominee

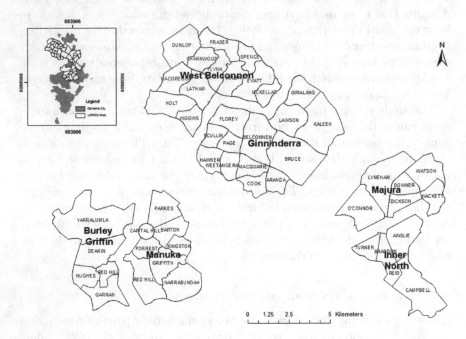

Figure 4.1 Six LAPACs shown in three different maps.

Source: Compiled from Department of Urban Services (2002)

2 one person is elected by business interests per suburb, three people in the Belconnen LAPAC areas
3 one person is nominated by the North Canberra Community Council, Manuka Residents Group or the Belconnen Community Council.

The Minister may appoint up to four additional persons, if necessary, to balance the areas of interest on the committee and cater for all sections of the community. The role of the committee is to comment on the broad planning directions for their communities. This is done through their development of Community Value Statements (CVSs), and by commenting on Territory Plan Variations, Section Master Plans and other important planning initiatives that may affect the planning and character of their neighborhoods (PALM, 2000a). One vital task is to develop and maintain CVSs for their areas. The CVSs define the community's aspirations for future development of their area and provide a framework for Committees to consider matters that are referred to them. The planning officers also use CVSs, while assessing the development applications. The committee members draft the CVSs in consultation with the community they represent; draft copies of the CVSs are provided to the members of the community for feedback and comment (PALM, 2000a).

Other aspects referred for comments include publicly notifiable applications for commercial, retail, industrial, community and multi-unit residential developments (excluding dual and triple occupancies), preliminary environmental assessments and other major urban developments affecting the area (PALM, 2000a). Committees meet once a month to discuss the issues referred to them and provide advice on planning and development matters. These meetings are open to the public. The Terms of Reference indicate that the committee's role is advisory, and members assist the planning agency for decision-making. The term of appointment to the committee is for two years, and the Minister may extend this for another term. Election of the committee members is open to all and eligibility of the members is minimal. People over eighteen years can be either residential or business representatives. If they can show proof of their residence or a business establishment or business interests in the suburbs covered by area. There is no other formal requirements to be nominated for the committee.

Every committee had a convener to chair the meeting, elected by the existing members. There was also a coordinator for all six cCommittees, who takes notes at meetings and organizes meetings and other related activities. The coordinator is a regular employee Planning Authority and is occasionally replaced by others. A planner (Technical Officer) attends every meeting to discuss planning procedures and represents the planning agency in the consultation process.

Besides providing comments through the established process on specific planning and development matters, the Minister for Planning meets with all conveners at least three times a year. General issues, policies and procedures are discussed in the meetings. Conveners are encouraged to solicit input from committee members for these meetings. The areas covered by each, all Committees are shown in Figure 4.1.

Selection of research participants

One of the most significant issues a researcher has to consider in designing a research project is the type and number of the respondents to be included in the study. Selecting participants in any qualitative or quantitative research is critically important. In focused interviews, participants generally are chosen on the basis of experience related to the research topic, which Swenson et al. (1992) refer to as 'purposeful sampling'. Since the aim of this book is to explore individual perceptions of the consultation process, the experience of the participants is required for making comments on the criteria it is supposed to evaluate through the Committee's consultation process. Planning professionals and others in the policy community attend the meetings to gain experience of the consultation process. This book approached Committee members who had considerable experience in the procedures and had been members of the committee for at least a year. This was ascertained through the coordinator who had contact lists of all members. Experienced members of the wider public, planning staff, the executive director of Planning Authority, spokespersons of political parties and the Minister for Planning were also selected as research participants in this book.

Lincoln and Guba (1985: 199) argue that there are some purposes behind all sampling procedures. This book used purposeful sampling to seek out experienced respondents to comment on the proposed criteria for evaluating the fairness and effectiveness of public participation in planning. This sampling technique allowed to select experienced members for an in-depth investigation of the consultation process and enabled them to select members who had either resigned from the committee or did not stand for in election for another term. The new members were not selected for the interview, as they had attended only a couple of meetings and had no or minimal experience of making comments on the overall consultation process. The enthusiastic wider public who often attend the meetings and have interests in planning matters were also formally interviewed in order to receive their feedback on satisfaction with the existing consultation process. The wider public are those who are actively involved with local planning issues through responding to development applications, attending Committee meetings, lobbying government and initiating petitions and protests in various community forums. A local newspaper, which has two versions, *The Belconnen Chronicle* and *The Southside Chronicle,* played an important role in selecting members of the wider public, including supplying the names and contact details of those actively involved in local planning issues. Some of these initial contacts became key informants, directing contact to others in the local community, who in turn directed the first author to other members of the

Table 4.1 Summary of interviews

The research participants	Number of participants interviewed
Committee members	62
Wider Public	12
Planning Staff	6
Minister for Planning	1
Executive Director of Planning Authority	1
Spokespersons of Political Parties	4
Total	86

wider public. This is often referred to as the snowball sampling technique. When using this technique, one should take care to avoid bias towards select-ing similar groups or one perspective. This was avoided by briefly surveying potential interviewees about their affiliation with the local community and their political involvements before selecting for interview. In most cases, the wider public were very vocal about their views and the views of people to whom they directed the first author. The continuing urban redevelopment program was a major issue for many of them.

Planning professionals were also interviewed. Interviews were held with the planners involved in the Committee's consultation process. The Executive Director of Planning Authority was interviewed separately. The Minister for Planning was interviewed just before the Territory election, when the Minister was the shadow minister for planning. The research participants are enumerated in Table 4.1.

4.3 Data collection

As discussed earlier, data collection for case studies can rely on many sources of evidence. Yin (1994: 78) identified six important sources for the case study inquiry: documentation, archival records, interviews, direct observation, participant-observation and physical artifacts. However, data collection for case studies is more complex than the processes used in other research strat-egies. Six sources can be divided into three broad techniques of qualita-tive data collection: document analysis, participant observation and focus interviewing.

4.3.1 Document analysis

Analysis using existing documents is a technique of data collection in qual-itative research, which can be combined with other data-gathering tech-niques such as interviews and observations (Yin, 2017). The collection of documentary data was important for this study, and consisted of community value statements, minutes of meetings, agendas, correspondence, newspaper articles on current development applications and resolutions of the ACT Legislative Assembly on the establishment of Planning Committees in 1995. Some documents such as meeting minutes, the Committee Guide and the Protocol were collected before the interviews were held.

Newspaper articles on the consultation and planning process published in *The Canberra Times, The Belconnen Chronicle* and *The Southside Chronicle* were also used in this book, and individual letters and emails sent to Planning Authority commenting on planning issues were referred to. Members' own letters, which they had retained and circulated at meetings, were collected from the Committee coordinator. Members of the wider public often attend the meetings with their written documents for making submissions on particular development applications, which were also collected from coordinator and used in this book.

However, the vast majority of documents used in this book were meeting agendas and minutes. The period from 2000 to the present was examined, as there were no archival records of pervious meeting minutes at Planning Authority or on the website. The first author took notes from the Minister's letters, normally sent to the committees, addressing many planning issues and indicating the government position. Most of the Committee members seemed happy to pass on all the documents in the white folder provided by planning authority for documents and correspondences. Some of them gave unrestricted access to the white folder where information on the consultation process described briefly.

These various archival records illustrate the kinds of community consultation currently practiced in ACT, and also given some indications of how the planning decisions have made. Most importantly, meeting minutes provide an opportunity to examine the comments of individual members and their agreements or disagreements on planning proposals.

4.3.2 Participant observation

Observation is one of the oldest methods of data collection; it means looking at an object or situation with a purpose (Sarantakos, 1998). Marshall and Rossman (1995) state that observation entails a process of systematic noting, recording events and behavior in the social setting chosen for the study.

Observations were carried out in this study by attending Committee meetings, planning workshops information rights and residents' association meetings. The author attended meetings occasionally to assist undergraduate students in carrying out their assignments. In one year, the author attended thirty-two meetings across all Planning Advisory Committees. Apart from the scheduled meetings, the first author also attended other consultation meetings of the Turner Residents Association, Old Narrabundah Community Council, Belconnen Community Council and Planning for ACT Together (PACTT). The first author also attended other consultation events regularly, including Neighborhood Planning Workshops and Workshops on Community Needs Assessment.

An observation checklist was used to record the important events, situations and phenomena observed during the meetings. This direct observation obviously helped to compare and verify the responses collected through interviews with all those involved in planning policy.

4.3.3 Interviewing

The interview is a method of collecting information about a person, upon a specific topic, through an organized oral communication between two people (Soubashi, 1998). Maccoby and Maccoby (1954: 499, cited in Dunn, 2000: 51) defined an interview as 'a face-to-face verbal exchange in which one person, the interviewer, attempts to elicit information or expressions of opinion or belief from another person'. An interview is a data gathering method in which there is a spoken exchange of information, and requires some form of direct access to the person being interviewed (Dunn, 2000).

There are three major forms of interviewing: structured, unstructured and focus or semi-structured interviews (Dunn, 2000). The structured interview is used for the measurement of phenomena such as behavior, attitudes, opinions, values, social characteristics, social conditions, relations and preferences of individuals on certain options. The information can be quantified on the basis of the coded answers and may produce data for descriptive and inferential statistical analysis (Fontana and Frey, 1994). The unstructured interview is used for revealing insights on a research topic or for explaining unexpected findings. It can be used either in exploring a broad problem, or identification of components of a general question (Soubashi, 1998). Another category of unstructured interview is the focused or semi-structured interview. Focused interviews are used either as part of a more quantitatively-oriented structured interviewing model, or qualitatively-oriented in-depth interviewing model (Minichiello et al., 1995). This form of interviewing is focused on issues that are central to the research objectives and questions, but the types of questioning and discussion allow for greater flexibility than does the survey-style interview (Minichiello et al., 1995). Thus, the semi-structured focus interview produces data on the insights, attitudes, perceptions and opinions of individuals. The primary objective of the focused interview is to elicit as complete report as possible of what was involved in the experience of the particular situation (Yin, 2017).

Given the strengths and weaknesses of various interviewing techniques, the focused interview is flexible in structuring both open-ended and close-ended questions and enables researchers to obtain both qualitative and quantitative information. This book used focused interview techniques for exploring the experiences of Planning Committee members and others in the planning policy community.

The focused interview

The focused interview differs in several respects from other types of research interview, which may appear similar at first glance. It has distinguishing characteristics (Merton et al., 1990) from other interviews in that:

> '...the persons interviewed are known to have been involved in a *particular situation;* they have worked in the same office building, lived in the same neighborhood, or taken part in an uncontrolled but observed social situation, such as a resident associations' meeting, a street demonstration, a community council meeting or a design review session on any development applications' (Zeisel, 1984: 139).

It is called 'focused interview' because it focuses on a specific topic, which respondents are asked to discuss, providing their views and opinions on research objectives (Sarantakos, 1998). Particularly, the interviewer introduces a stimulus related to an issue with which participants are familiar, and discusses these issues with them. In a more general sense, it is a semi-structured interview (Sarantakos, 1998). In brief, the *focus interview* is focused on the subjective experiences of persons involved in a particular situation (Zeisel, 1984).

The focused interview can be used either with individuals or in a group setting to obtain in-depth information on how people define and analyze a concrete situation? what they consider important about it? what effects they intend their actions to have? and how they feel about the situation?(Zeisel, 1984). The individual interview involves an interviewer asking questions of a single individual in a quiet and private setting, and the focus group involves a collection of individuals answering the same questions together as a group (Soubashi, 1998). Focus groups are useful to reveal the opinions and thoughts of participants more effectively, when a group is socially and intellectually homogeneous (Merton et al., 1990). Given the strength of the focus group interview, this book conducted formal individual semi-structured focus interviews instead of group interviews, because it was observed that some members in every committee usually dominate the discussion and other members remain silent unless they are asked to respond. Merton et al. (1990: 148) refer to this as the 'leader effect': in most groups of people one or two inevitably emerge as louder, more dominant or more opinionated. Therefore, the individual interview is easier to control, compared to the focus group interview.

Rationale of focus interview for the study

Evaluation of the public participation process relies heavily on the perceptions, attitudes, feelings, beliefs, experiences and reactions of the participants. Minichiello et al. (1995) note that an individual's perception is influenced by past experiences, values, moods, social circumstances and expectations. In other words, an individual's understanding and belief about a situation are shaped by a variety of factors and may vary over time. Consequently, two individuals viewing the same situation may perceive it differently. Minichiello et al. (1995) also point out that individual perceptions cannot be measured directly, rather inferred from observing behavior or listening to what people say, so researchers must often rely upon interviews for the collection of necessary data about perceptions and viewpoints. In order to get data, whether qualitative or quantitative, the interview method is perhaps one of the most common and effective ways to understand people's attitudes and perceptions (Fontana and Frey, 1994).

The survey technique could have been used, asking open-ended or structured questions about the form, functions, fairness and effectiveness of a consultation process; but this would not have allowed for interaction between the respondents to clarify and explain questions and responses (Kvale, 1996). Thus, focus interviews were used in this book to obtain interview data from members and others concerned with planning policy who have long been involved in consultation on planning and development in Canberra.

For these reasons, interviews were held with Committee members and others in the planning policy community, with a number of predetermined criteria and question sets, but allowing sufficient time and flexibility to digress or probe beyond the answers to the standard questions. The purpose of the questions was to explore the perceptions on the proposed evaluative framework using two meta-criteria: fairness and effectiveness in the public participation process. This type of interview enables a researcher and the respondent to establish interpersonal relationships to explore sensitive and unspoken issues, and address additional relevant issues as they may arise during the interview (Fontana and Frey, 1994).

In the 'fairness' section in the interview schedule, respondents were asked questions on the consultation process, which they interpreted giving examples from their own experience: more interestingly, they discussed their roles in decision-making. They were also asked to justify the present form of advisory committee and clarify its fairness process, and to determine perceptions as to whether the advisory committee was an effective system of public participation.

In the 'effectiveness' section, respondents were asked to discuss the way in which the consultation process operates. They were encouraged to describe the operational or functional problems faced, and the relationships among Committees, Planning Authority and others in the planning policy community, as part of the larger planning process. All respondents were asked to provide their own perceptions on problems of the current Committee and the merits, if any, of their recommended process. To develop an understanding of an ideal consultation situation, those interviewed were asked their views of the categories that would refer to their past experience. This information was presented in many forms in order to recommend new criteria for evaluating public participation in planning.

Form of the semi-structured focus interview

The semi-structured focus interviews involved a combination of semi-structured and unstructured interviews. Questions were developed for each of the criteria selected for developing a theoretical framework as discussed in Chapter 3. An interview guide was purposefully designed to address the proposed criteria of fairness and effectiveness in public participation, but the form of the interview followed the basic focus interview strategies proposed in the methodology literature (Zeisel, 1984; Merton, et al., 1990; Foddy, 1993; Fontana and Frey, 1994). Before respondents answered questions, the objectives of the research were stated to them, and the first author explained the criteria of fairness and effectiveness to avoid the differences in meaning between first author and respondents, who were then asked to answer questions on the sub-criteria and other related questions. A drawback in using this technique is that those interviewing may comment on altogether different criteria. In such cases, the probing technique is advocated to bring the discussion back to the main criterion. After discussing many sub-criteria, the first author used the interview guide to incorporate them into the main criterion that discussed before.

The first author obtained contact details of both former and current members. Current members were initially approached for an interview just after the Committee meeting, but most were contacted by telephone and asked to set a time and place for the interview. Most of the interviews with residents and community representatives were held in their homes, but a few residents and business members preferred to meet at their offices and local coffee shops; some came in the first author's office.

The initial interviews were held face-to-face, but some respondents were subsequently telephoned, when required to clarify comments and opinions on specific criteria. Interviews were continued between one and two hours, while most were completed in one hour. Two exceptionally long interviews with Committee conveners took three hours to complete. Follow up interviews in person or by telephone lasted from ten minutes to one hour.

As the purpose of the interviews was to obtain an understanding of attitudes and perceptions, the order of the questions was not important. During each interview, there were opportunities for open discussion on many different issues and concerns related to consultation and the decision-making process. The focus interview allows a researcher to be flexible and change the question order and the wording of the criteria to be covered (Kvale, 1996). Sometimes the first author asked questions at appropriate times during the conversation; in this way, some new issues and concerns were introduced into the analysis and evaluation of each criterion.

The relationship between the interviewer and respondents is often critical to the collection of opinions and perceptions – if they know each other, the respondent is likely to be communicative (Dunn, 2000). Goode and Hatt (1982, cited in Dunn, 2000), warned that interviewers should remain detached and aloof from the person interviewed but the interviewer must introduce at the beginning of the conversation. The first author described his role and purpose of interviews, and gave an assurance that respondents' anonymity would be preserved. However, some business representatives held the impression that this book would only focus on the resident and community views and would undermine the views of development proponents, probably because the first author was introduced by conveners who often represented either resident or community groups. The businesspeople were reassured by the information that the study would accommodate the concerns of the whole planning policy community, not just the residents.

Recording interviews

The entire interview was tape-recorded, although some notes were also taken. Tape-recording and note taking are the two main techniques for recording interviews (Dunn, 2000). This book mostly used tape-recording while notes were only taken to identify the criteria a respondent was discussing. Before commencing interviews, the consent was sought to use the tape recorder. Some respondents were reluctant to give permission so in such cases only notes were taken. Others agreed to the use of the recorder and even allowed their names to be used, but names were eventually changed, where appropriate, to ensure respondents' anonymity. Sometimes, respondents asked for the recorder to be stopped when they were explaining some important events, which they termed 'off the record'.

Both tape recording and note taking have advantages and disadvantages. A recorder can record whole conversations, while note taking requires shorthand-writing skills to produce verbatim records. However, the primary aim in note taking is to capture the gist of what is said (Dunn, 2000), so notes were taken only for this purpose and to identify the criteria. An additional advantage of tape recording is that it allows a researcher more time to organize the next question and to maintain the flow of conversation without any break.

Ethical considerations

This book is based on data obtained from interviews that involved personal experiences and opinions of a particular process. The opinions were often expressed emotionally. And since the Committee members are known in the ACT, the confidentiality and anonymity of their comments and individual identification was a major concern. Kvale (1996) has identified three ethical guidelines for maintaining confidentiality and anonymity in human research, which are informed consent, confidentiality and consequences. Informed consent entails informing the research subjects about the overall purpose of the study, possible risks and benefits of participation in the study and the right to withdraw at any time during the interview. Confidentiality in research implies that it will maintain anonymity. The consequences of an interview study need to be addressed with respect to possible harm to the subjects as well as the expected benefits of participating in the study. These three ethical considerations had always been maintained throughout the interviews. All interviewees were briefed about the purpose of the study and asked whether they wanted their comments to remain confidential. Although most were happy for their actual names and comments to be recorded, a few did not want to share their names. However, it was decided to maintain the anonymity of all interviews.

Group interviews

A more formal approach to group interviews is known as the focus group (Morgan, 1997). The focus interviewing technique is as useful with groups as with individual respondents (Zeisel, 1984). Although this research mainly carried out individual interviews, but some informal group interviews were held with Committee members and members of the wider public without formal arrangements of date and time. The later occurred after the close of meetings when members sat around a table outside the meeting room and discussed the issues covered in the meetings. For example, when the Committee members discussed the responsibility of Planning Authority staff to give them adequate feedback, the participants in group interviews were asked to give examples of delayed feedback and insufficiency of information. All the members round the table made comments on the overall feedback system in the consultation process, which were recorded. Some planning staff were interviewed individually and some as a group; where the group interviews were necessitated by the time constrains for planning staff. They were agreed to interview together, but agreements and disagreements were recorded and subsequently analyzed.

Closing the interview

Some preparations are required for the closure of an interview; otherwise the ending can be clumsy (Dunn, 2000). The ideal is 'to provide a summary, set post-interview goals, and exit gracefully' (Donaghy, 1984: 11). The interviews conducted for this book were ended in several ways. When respondents had agreed upon a specific time for completion, the author tried ways of concluding the conversation in a timely fashion. Some respondents appeared to want to talk on other topics at a later time, which was done if follow-up interviews were required. On the other hand, respondents were carefully observed to see whether they were eager to continue the discussion. Looking at the clock or yawning indicated that it was time to end the interview. It is important that the researcher should express not only thanks but also satisfaction with the material collected (Dunn, 2000), and also assure that anonymity and confidentiality would be maintained.

4.4 Data analysis

There are many ways to analyze qualitative data. Creswell (1998) describes approaches to qualitative data analysis within five traditions of inquiry: biography, phenomenology, ethnography, grounded theory and case study. In biography, 'a researcher begins analysis by identifying an objective set of experiences in the subject's life...and develops a chronology of the individual's life' (Creswell, 1998: 146), and identifies factors that have shaped the individual life. In phenomenological analysis, 'the researcher begins with a full description of own experiences of the phenomenon' (Creswell, 1998: 147), describes individual experience with the topic that emphasizes on the statements made by individuals on a particular process. The researcher then develops a list of non-repetitive and non-overlapping statements, and finally, 'constructs an overall description of the meaning and the essence of the experience' (Creswell, 1998: 155). Like phenomenology, 'grounded theory provides a procedure for developing categories of information (open coding), interconnecting the categories (axial coding), building a 'story' that connects the categories (selective coding), and ending with a discursive set of theoretical propositions' (Corbin and Strauss, 1990: 7).

Ethnographic research is based on three aspects of data transformation for interpretation: description, analysis and interpretation of the culture-sharing group (Wolcott, 1994). The description may be analyzed by presenting information in chronological order or narrator order, and the analysis involves highlighting specific material introduced in the descriptive phase through tables, charts, figures and diagrams. The interpretation of the culture-sharing group is also a step in data transformation. In this context, the 'researcher draws inferences from the data or turns to theory to provide structure for his or her interpretations' (Wolcott, 1994: 44). As with ethnography, 'case study analysis consists of making a detailed description of the case and its setting' (Creswell, 1998: 153), based on multiple sources of data such as 'documents, archival records, interviews, direct observation, participant-observation and physical artifacts' (Yin, 1994: 78). Stake (1995) analyzes four forms of data analysis and interpretation in case study research, namely categorical aggregation, direct interpretation, matching patterns and naturalistic generalization. In categorical aggregation, researchers collect examples from the data sources, in the hope of finding issue-related meanings. In direct interpretation, the researcher looks at a single example and draws meaning from it without looking at multiple examples. In pattern matching, the researcher looks for similarities or dissimilarities between two or more categories. Finally, in naturalistic generalization, the researcher analyzes what the people can learn from the case study.

All the approaches mentioned above use the inductive data analysis method. Inductive analysis means that the patterns, text, themes and categories of analysis come from the data. They emerge from the data rather than being imposed on them from external sources based on theory (Patton, 1990). Most of the data collected in this study consists of texts. All sources yielded mainly textual data, in the form of literature on public participation in planning, planning reports from Planning Authority, written submissions and transcribed interview data. The main sources were focus interviews and observational notes taken during the Committee meetings and other planning workshops.

The common analyzes of focus interview results involve transcribing audiotapes and field notes (Stewart and Shamdasani, 1990). All transcribed data are organized under various headings, which should correspond broadly to the questions posed during the discussion (Krueger, 1994). All interviews carried out for this book were transcribed and coded according to the evaluative criteria discussed in Chapter 3. Once the entire interviews were completed and the supporting notes and documents had been obtained and reviewed, the criteria for the evaluative framework were applied. This involved the systematic identification of essential issues in the interview data, and relating them to the appropriate meta-criteria of fairness and effectiveness. Kvale (1996: 189) notes that separating data in 'qualitative research consists of four different elements or parts, including categorization, condensation, structuring, and interpretation of meaning'. He proposed analyzing qualitative data through six steps, which offer guidelines for qualitative researchers to carry out qualitative research. The application of the evaluative framework used in this book followed Kvale's (1996: 189) six steps, which are listed below:

1 Subjects describe spontaneously what they experience, feel and do in relation to a topic.
2 Subjects themselves discover new meaning in what they experience and do. This process is free of interpretation by the researcher.
3 The researcher condenses and interprets the meaning of what the interviewee describes and 'sends' the meaning back. The conversation between researcher and interviewee continues until the interpretation is clear.
4 The researcher interprets the transcribed interview either alone or with other researchers. This process involves structuring information according to topics, clarifying material by eliminating superfluous information and developing the meaning of the interviews.
5 Interviewees are re-interviewed for further clarification.
6 Future interviews are altered based on new issues arising from earlier interviews.

In a broader sense, the aim of this book is to evaluate public participation in planning using the proposed criteria of fairness and effectiveness. According to Zeisel (1984), research that aims to explore the perceptions of individuals or groups in a particular situation, depicting the experiences in situational analysis, requires exploratory research with description and interpretative analysis. Using Kvale's approach, the interview data, notes and supporting documents were combined to analyze the case study of Planning Advisory Committees. However, steps three and four of Kvale (1996) above, which require condensation, categorization and structuring the information of textual data, have been analyzed using the process of de-contextualization and re-contextualization of data proposed by Tesch (1990: 142–145).

4.4.1 Data analysis procedures: De-contextualization and re-contextualization

There may be several components in the discussion of the plan for analyzing the data (Creswell, 1994). The process of data analysis is diverse, so there is no 'right way' (Tesch, 1990). Data analysis requires that researchers should be comfortable with developing criteria or base lines to study their own objectives (Patton, 1990), and that they should be open to possibilities and see contrary or alternative explanations for the findings (Creswell, 1994). Patton (1990) notes that qualitative methods generate many data, which researchers try to reduce to certain patterns, categories or themes, and then interpret the data by using some schema. Tesch (1990) calls this process of organizing textual data as 'de-contextualization' and 're-contextualization'. He proposes eight steps to systematically organize textual data. This book used these eight steps (Tesch, 1990: 142–145):

1 Acquire a sense of segment of the whole text through transcribing the data from various sources.
2 Transcribe the representative interviews from all stakeholders and recorded topic descriptions above the text segments (paragraphs).
3 After examining all the recorded topics, create a master file, grouped the topic into evaluative criteria as proposed in Chapter 3, and divide them into the categories 'most related', 'moderately related' and 'loosely related'.
4 Abbreviate the topics as coded and write the codes next to the segments of text.
5 Use master list of codes and add other transcribed documents (meeting notes, transcription of minutes, and documents notes).
6 Create a new abbreviation list for each category and alphabetize these codes.
7 Assemble the data material belonging to each category in one place and perform a preliminary analysis.
8 Record the topics when necessary.

4.4.2 Organizing textual data

The first step involved transcribing the audiotapes, which produced 1169 pages of textual data. All transcribed data were divided into various text segments such as lines, paragraphs and pages according to the criteria proposed to evaluate the public participation process, which Tesch describes as 'de-contextualization' (Tesch, 1990: 143). These textual data were merged with other transcribed documents and field notes. Complete transcriptions of data were not included in this book, but relevant brief comments have been inserted. Each line, paragraph and page was related to the proposed criteria, the paragraphs in particular, were given code names corresponding to each criterion. Nodes were placed in each paragraph to identify the key quotations used in written materials. This helps to distinguish and compare the diverse opinions of the planning policy community in order to identify their agreements or disagreements on many planning issues. The transcription was then reread, and the appropriate code written in the right margin highlighting words, sentences, and paragraphs related to each criterion.

Once all the text segments were identified in the right margin, a master list was created. This was helpful in identifying the text segments and various coded texts that were divided into the fairness and effectiveness criteria. Each criterion in the text segment was given an abbreviated code and all codes were applied to each of the remaining text segments where appropriate. This step is referred to as 'tagging' or 'coding'. This step was followed for each interview with the planning policy community and merged with data from other sources such as documents and notes taken in meetings and workshops.

Once transcriptions were completed, a preliminary heading was given to each criterion, providing an understanding of the text units within the context of the entire interviews. An important element of data collection and analysis was the tagging and regrouping of individual submissions, the Minister's letters, and other memos. All memos were manually coded and merged with those under various categories.

4.4.3 *Computer software for qualitative data analysis*

The assembled, i.e., the re-contextualization categories were imported into Nvivo, a computer program specially designed for qualitative data analysis. Other software such as NUDIST, Ethnograph and ATLAS/ti, can also be used, and are particularly helpful when there is a large amount of transcribed material to be analyzed (Cameron, 2000). Nvivo is new-generation software designed for research that needs combining the subtle coding with qualitative linking, shaping and modelling (Bazeley and Richards, 2000). It allows researchers to input field notes and recorded comments into the appropriate categories for organizing interview data into a rich text format file with necessary coding into the text by placing nodes in every paragraph (Bazeley and Richards, 2000).

The Nvivo software helped to sort the text segments, identify nodes in the paragraphs and divide the paragraphs according to the criteria adopted for fairness and effectiveness analysis. It was useful in searching for key criteria within a large database of focus interview documents. Nvivo has tools for handing all these qualitative data, and supporting techniques to organize textual data derived from various sources. Although Nvivo is capable of coding all text segments, the author identified the topics and coded the text by hand, which gave the author a control to identify text segments and organize transcription into the proposed criteria for the evaluation.

4.5 Reliability and validity of the research

Without the use of trustworthy methods, research becomes fiction and loses its utility (Morse et al., 2002). In a broader sense, reliability and validity address issues about the quality of data and appropriateness of methods used in conducting a research project (Mason, 1996). Reliability is one of the central concepts in assessing the quality and trustworthiness of the research. Reliability 'refers to the degree of consistency with which instances are assigned to the same category by different observers or by the same observer on different occasions' (Hammersley, 1992: 67). Validity addresses whether the research explains or evaluates what has been said would be explained and evaluated, 1992). Validity is another concept in assessing the quality and rigour of a research. (1992: 57) claims that validity means truth. For him, validity is interpreted as the extent to which an account accurately represents the social phenomena to which it refers. This methodological rigour is significant in a study using qualitative design. Patton (1990: 462) notes that 'qualitative research has an obligation to be methodological in reporting sufficient details of data collection and the process of analysis to permit others to judge the quality of the resulting product'.

However, reliability and validity are particularly difficult to achieve in unstructured interviews (Donaghy, 1984: 233), because human experiences are unique, particularized and not always verifiable (Hall and Stevens, 1991). Therefore, there are concerns to the rigour of qualitative inquiry (Morse et al., 2002). In the absence of quantitative numbers and p values, qualitative inquiry results in a lack of confidence from both inside and outside the field. However, a number of qualitative researchers (Leining, 1994; Altheide and Johnson, 1998) argue that reliability and validity are relevant to quantitative inquiry, but not to qualitative inquiry. Some have suggested adopting new criteria for determining reliability and validity, and thus ensuring trustworthiness in qualitative inquiry (Lincoln and Guba, 1985; Leining, 1994; Rubin and Rubin, 1995).

Guba and Lincoln (1994) replaced reliability and validity with the parallel concept of 'trustworthiness' for qualitative inquiry, involving four aspects: credibility, transferability, dependability and confirmability. They see these categories as broadly equivalent to the concepts of validity, generalizability, reliability and objectivity that have been used to evaluate the quality of quantitative research endeavors.

4.5.1 Credibility

Credibility involves establishing that the results of qualitative research are credible from the perspective of the participants in the research. Sandelowski (1986) follow Lincoln and Guba's (1985) criteria and comment that credibility in research could be claimed when it describes the interpretation of people's experiences and their perceptions on a particular issue (Sandelowski, 1986). Lincoln and Guba (1985: 296) argue that the qualitative research inquiry should be carried out in such a manner that the probability of findings should be trustworthy, and that it should provide ample opportunities to the respondents to look at and evaluate the research findings. Criteria for creditability of this research consists of prolonged engagement, topical observation, triangulation, referential adequacy and member checks.

Prolonged engagement of the researcher for a long time is beneficial for prior familiarization with the research topics and understanding the nature of the issues to be discussed and explained. Through this, a researcher can understand the authenticity of the responses given by the respondents and build trust. Since the first author had been attending so many committee meetings for more than a year, his familiarity with the consultation process was particularly helpful in understanding the authenticity and establishing trust among all planning policy community about the nature of this study.

Lincoln and Guba (1985: 304) noted that the aim of *topical observation* is to clearly identify the issues and criteria that it is supposed to analyze. The first author mostly took extensive notes during the meetings, which were verified with the interview data. This opportunity enabled to identify relationships, conflicts and trusts among all those involved in planning policy.

The use of several data sources and different methods is called *triangulation*. The idea behind triangulation is that the more agreement of different data sources on a particular issue, the more reliable the interpretation of the data (Flick, 1998). Qualitative methods usually research a question through several methods. It is not unusual to use a combination of documentary analysis, participant observation and interviews in qualitative research. The use of multiple methods and various data sources in research increases its reliability. This book used the multi-methods approach, which provides effective cross-checking of information. Another way of cross-checking information was interviewing both Committee members and planning staff involved with the consultation process. A final cross-check was ensured by interviewing former Committee members, retired Town Planners and the planning spokespersons of four political parties. Data triangulation occurred because of using focus interviews, review of documents and direct observation during the meetings.

Referential adequacy ensures the reliability of the collected data. To achieve referential adequacy there was a need for visiting the development sites and neighborhoods, which have high re-development pressure. Few such neighborhoods are the Kingston Foreshore Development in Kingston, development activities along Northbourne Avenue and Barry Drive and the Hiagh Park in the inner city area.

4.5.2 Transferability

Sandelowski (1986) commented that the essence of a qualitative inquiry should be meaningful directions and applicability to other situations. Lincoln and Guba (1985: 316) note that 'transferability in qualitative inquiry refers to the usefulness of utilizing the study process and its results in the context of another time and place'. This book is confined to the evaluation of the consultation process through Planning Advisory Committees and proposes a theoretical framework to evaluate the fairness and effectiveness of public participation in planning. This evaluative framework can be applied to other public participation processes in general and Planning Advisory Committees in particular. Moreover, the readers can determine the applicability of the findings and their relevance to other contexts.

4.5.3 Dependability

Dependability is ascertained by examining the methodological and analytical decisions made by the researcher during the study (Hall and Stevens, 1991). This may require determining whether the decisions made are consistent with their surrounding circumstances. It also warrants an assessment if the existing data support the interpretation and recommendations of the researcher. For a research to be dependable, its data collection, sampling, analysis and result dissemination should documented by their related rationale, outcome and evaluation of the actions (Hall and Stevens, 1991: 19).

4.5.4 Confirmability

Qualitative research tends to assume that each researcher brings a unique perspective to the study. Confirmability refers to the degree to which the results can be confirmed or corroborated by others (Lincoln and Guba, 1985). There are several strategies for enhancing confirmability. The researcher can document the procedures for checking and rechecking the data throughout the study. This is achieved through the researcher's audit trail, which allows the researcher to track the decisions made and steps taken in the study. An audit trail entails keeping a research journal that includes original data (such as audiotapes, transcripts and meeting notes), early data interpretation or analysis, and guides for Committee members. All details were recorded in a journal and inserted in this book where appropriate.

4.6 Data presentation and reporting

When preparing the written report, a researcher considers the style of reporting. Focus interview reports have traditionally been presented in a narrative style (Krueger, 1994), which uses complete sentences and is augmented with quotations from the individual or group interviews. An alternative is the 'bulleted' or outlined report, which uses key words and phrases to highlight the critical points, but this style is normally restricted to short report writing. This book used the narrative style. There are three narrative styles in focus interview reporting (Krueger, 1994): the first consists of the quotations or ideas followed by all participants' comments. The second style is a summary description with illustrative quotations followed by an interpretation. This book followed the third narrative style, which is replete with statements and followed by the researcher's interpretations of those statements.

One important aspect of using the narrative is finding a balance between direct quotations and the researcher's summary and interpretation of the discussion (Cameron, 2000). Morgan (1997) notes that when too many quotations are included the materials can seem repetitive or chaotic, while too few quotations can mean that the vitality of the interaction between participants is lost to the readers. Therefore, Morgan (1997) recommends that the researcher should aim to connect the readers and the original participants through 'well-chosen' quotations, which this book has done.

4.7 Chapter summary

Care has been taken to use appropriate methods to evaluate the criteria of fairness and effectiveness in public participation. Understanding people's perceptions and differing opinions on the same issue requires qualitative inquiry. Using the semi-structured focus interviewing technique is a relevant way to gather the attitudes of people involved in planning policy and their perceptions on the overall consultation process. The flexibility of focus interviews allows the respondents to elaborately discuss issues of particular criteria that they may consider pertinent. In addition, field notes collected during the meeting, documents such as publications on community consultation, and meeting minutes have given credibility to the research methods used for this book.

There is much literature about the criteria of effectiveness in public participation (Webler, 1995), however, most of it related to environmental decision-making (Day, 1997). Very little has been written on evaluating public participation in planning, so, the aim of this book is to develop a conceptual model that can be applied in planning research. The Planning Advisory Committee case study has provided some important criteria that need to be addressed in evaluating fairness and effectiveness. The focus interviews yielded many comments on these criteria, which were organized using the computer software Nvivo. The presentation of data followed both the narrative and the interpretative style. The results are discussed in the following three chapters.

5 Evaluating fairness in public participation process in urban planning

5.1 Introduction

The aim of this chapter is to evaluate and coherently discuss the criteria of fairness. Research on fairness demonstrates that people's satisfaction with a decision largely rests on whether or not the consultation process has been carried out through fair procedures (Lind and Tyler, 1988). If people perceive the decision-making procedures as fair, they are more likely to perceive the decisions as fair (Lind and Tyler, 1988). Perceptions of fairness are often closely correlated with people's satisfaction with a consultation procedure and their perceptions of decision outcomes. Thus, it is essential for public participation efforts to be conducted fairly in the eyes of citizens and consider their evaluations of a consultation process (Smith and McDonuch, 2001). Although public participation literature acknowledges the importance of evaluating fairness, researchers have drawn a variety of conclusions about the criteria people use in deciding whether a procedure is fair (Lauber and Knuth, 1999).

For this book, respondents were asked to evaluate their public participation experiences. Various issues of fairness were analyzed in depth to help determine how respondents themselves conceptualize a fair process for planning decisions. The evaluation is derived from data acquired through focus interviews, observation and supporting documents, as previously described. This chapter is organized on the basis of the objective of evaluating the Planning Advisory Committees (PACs) through the proposed meta-criteria of fairness of the public participation process in planning, with specific reference to the method of the Planning Advisory Committee (PAC). The essential criteria discussed are adequate opportunity, early involvement, representation and agenda setting and feedback. The analysis of these fairness criteria can further be used to formulate or to modify people's satisfaction with the consultation process.

5.2 Criteria for evaluating fairness in public participation

5.2.1 Evaluating fairness: Adequate opportunity

An adequate opportunity criterion has many factors, of which four are pertinent for the discussion of fairness in a PAC: equal capacity to participate

DOI: 10.4324/9781003122111-5

meaningfully in the discussion process, equal access to pertinent planning information, equal power of influence on the planning decisions and equal opportunity to participate in the discussion. However, every criterion has its own sub-criteria, which are systematically discussed below.

Equal capacity to participate meaningfully

To participate in the discussion process meaningfully, participants require adequate access to planning information that is easily understandable by the planning policy community. Webler (1995) defined this as competent understanding by individuals of issues required for the discussion. However, the authority that conducts the consultation should provide relevant and adequate information to the participants, and it is essential that adequate time should be provided for making informed comments on planning and development matters (Dandekar, 1982). Therefore, dissemination of relevant planning information and communication of this information to the participants are necessary for the development and implementation of planning policies (Dickinson, 1999). Access to relevant planning information is also increasingly recognized as essential for informed planning decisions. But what type of information is provided to the planning policy community? This question concerns the relevance of information provided to participants for their comments and suggestions (Kaufman, 1999). In this regard, there were some problems related to 'inadequate relevant information' and 'too much disorganized information' provided to the Committee members for their comments on development applications.

The Planning Committee reviews and provides input on all formal development applications, Master Plans, Group Centre Redevelopment Plans and other planning and design matters referred by the planning authority. The Committee members, with their relative understanding of planning and design matters, provide a great deal of community concerns on proposed development applications, which is their main role in the meetings discussing planning proposals. To fulfil this role, committee members must review a large amount of planning and design information before discussion can proceed. The documents may include all background and preliminary studies, previous applications, maps, information on zoning, sections of the relevant *Land (Planning and Environment) Act 1991* (now *Planning and Development Act 2007*), the existing Territory Plan and the guidelines for achieving High Quality Sustainable Development. It is important for the planning authority to forward all relevant information well in advance of the meeting date to the Committee members, allowing them ample time to review development applications. However, the Committee members interviewed were found to be dissatisfied with the time allocation for reviewing planning documents, which they believed was insufficient. Most members believe that they are normally given documents too late, sometimes just a day before the meeting. Delay in forwarding the relevant information to the members does not give them enough time to review planning proposals for discussion at the next scheduled meeting, so they are unable to participate meaningfully in the discussion. Similarly, time allocated for written submissions on specific development applications is also insufficient; in some instances, such submissions were forwarded to the members only on the evening before the scheduled meeting day. In such cases, Committee members cannot examine the details of development applications to make comments that reflect community concerns about the planning decisions.

Some members also expressed dissatisfaction that they were not always given all the information necessary for making informed comments and had to request additional information from the developers or planning staff. Most of the Committee members believe that planning staff control the level of input provided by the Committee members and, therefore, the flow of information. The Committee often finds it difficult to determine the staff position on proposals, whether all information has been provided to the Committee, as the developers sometimes attend without providing necessary information to the Committee members or to the concerned Technical Officers. Members become confused as to whom they should approach for necessary information required for the evaluation of a development application.

However, the protocol of PAC (PALM, 2000b) indicates that every Committee must operate within a timeframe in order to approve a development application. A number of Committee members find that the preparation for some specific meetings is very time-consuming, especially when they are required to review background materials for the Master Plan and Group Centre Re-development Program; so they find it difficult to prepare themselves adequately on the complex development applications, which they feel require ample time before the Committee meeting. Unfortunately, the planning authority allocates same timeline regardless of whether the development applications are simple or complex. Most of the members find that this process prevents them from participating meaningfully in the meetings.

On the other hand, the protocol also states that the main objective of the Committee is not to delay the approval of development applications (DAs), but to involve the community for their input at the earliest possible time in order to incorporate their comments into the planning decisions. Generally, there is a time limit to get DA approved after formal submission to the planning authority, so Technical Officers and Committee members must respond within a limited timeframe. It is observed that time for reviewing DAs is not enough to prepare adequately for an informed discussion. Only the business representatives on the Committee were found to be satisfied with the time allocated for reviewing development applications. Complaints about insufficient time given to resident and community groups are questioned by the business representatives, who believe that residents and community representative do not understand planning procedures and development approval. Those members dissatisfied with insufficient time allocation for comment categorically expressed their concern about business representation on the

Planning Committee. The residents and community representatives were mostly dissatisfied and opposed to business representation. A resident representative illustrated this issue directly:

> Business reps [representatives] who are *de facto* representatives of developers know all the rules and regulations, therefore, they don't require sufficient time to make comments. But we need time to go through detailed plans.

But this does not happen in Planning Authority's consultation process for allocating time to all Committee members. Authority normally gives very little time to read and to make comments on the DA just a day before the scheduled meeting, or even during the meetings when member do not have a chance to glimpse the documents. A resident member said:

> We have particular concerns that the developer knows that the decision of DA is going to occur between or just before the meetings. DA can be lodged just after one meeting and decisions made before the next meeting. So, there is not enough time to make comments on DA most of the time.

Another member, who had long been involved with a Committee, made a similar comment on time allocation:

> Generally, there is a time problem. DA copy needs to be sent to every member of the committee. But it is not happening. If we are going to make a decision on a development application, every member needs to be able to have the opportunity to look at the plan before the meeting. What they [Authority and developers] are doing now is sending a copy to the convenor, a copy to the representative of that particular suburb, not everyone. Only the members who belong to that particular suburb have the opportunity to see it before the meeting. You cannot make any competent judgement without a copy of the DA before the meetings.

Equal access to the pertinent information

Access to the various pertinent documents is an important criterion for evaluating the fairness of a consultation process. The Advisory Committee and others involved in planning decisions should have equal access to the sources for commonly agreed-on standard definitions of the continuing development process (Webler and Tuler, 2000). All participants in the planning process should also have equal access to available and systematic knowledge[1] (Renn et al., 1995).

Residents and community representatives often complained that they were given inadequate information. The white folder provided to the Committee members at the beginning of their tenure is messy. Most of the members commented that the folder is not organized in such a way that they can find the relevant planning laws and design guidelines, which are basic to comments on a development application. Often, they had to ask the Technical Officer or coordinator to provide relevant information. A resident member commented:

New members would not understand what is contained in the white folder. It is not well organized and has no community profile. Most importantly, it does not have basic information with an executive summary, so that a new member could understand at least the basis of development approval.

All planning participants should be given planning documents written in accessible language. The quality of planning information provided to the Committee members is important for making informed comments on planning and development matters, but members interviewed were dissatisfied with the quality of planning information provided to them. They were asked to evaluate the quality of information given to them in the form of development applications and Master Plans. The residents and community representatives expressed dissatisfaction with the contents and the information structure. They felt that most of the documents used planning terms and many technical words, which they believed a layperson would not understand clearly. Thus, a major constraint to understanding the planning information was the language used in documents provided to members with no prior knowledge of planning and design matters. Community groups also complained that documents were not written in community-accessible language, but in planning and design jargon, which Committee members believe is difficult to understand. They feel that this hinders their meaningful participation in the discussion. A community representative on the Committee commented:

Well, the architect and planner can understand all the documents, it is not written for us.

Another resident member, by profession an architect, commented that the amount of information was sufficient, but the language used to communicate the information made it inaccessible to some Committee members. He added:

There is always complaint about the information that is provided. And, this I would tell you; it is not everybody that understands it. I mean, me as architect planner, I have… a way of understanding planning and high-quality sustainable design concepts, it is not everybody that understands this. ACTCode 2, criteria of high-quality sustainable design…

not everybody understands the objectives of this type of language. ...But the basic information is available to everyone. The question is how is it [information and documents] structured?

The interview data indicate the level of members' perception of the language used in the planning and design documents. There was no unconditional support for the statement that planning documents should be in accessible language. Rather a clear demarcation was noted, which is to say that some documents are comprehensive and need a planning professional to understand them. The new members find difficulties in understanding the development applications and other planning documents, which results in meetings being dominated by old members while the others are marginalized.

Equal power of influence

The interview data indicate that two groups have strong influence on the decision-making process: senior planners or assessors and development proponents. The Committee members interviewed feel that Technical Officers have greater opportunities than others to influence the DA assessing team within the planning authority. The wider public also believe that the planning authority depends on the views of Technical Officers in making final decisions, except for other planning issues with political interests. Some Committee members feel that much of the policy process takes place behind closed doors in discussions between Technical Officers and developers. To some extent, Planning Authority's Executive Director and other planning staff interviewed also acknowledged that Technical Officers played an influential role in final decisions. Some members who regularly attend the meetings believe that it is not the Technical Officers or assessing group, but the developers, with good professional interactions with the higher authority, are mainly the influential actors in the decision-making process. Both parties interact and have influence at the beginning of their DA process, which has been criticized by many members, who term it an 'unholy alliance' between planning staff and proponents. This is reflected in a comment made by a West Belconnen PAC member:

A developer in the first instance goes to see them [planning staff]. He [developer] comes up with the proposal and staff gives some guidelines and says: *well, this is what you have to do, and comply with until or before you go to see the people.* So, they have developed some strategies where they have to say, *you have to inform the public and let us know.* What is happening is, planners and the developers come hand-in-hand and give the impression that planner is in the pocket of the developers. And planner say *that is what we are doing for the community, this is good for you, you should have it,* as if they knew what we wanted.

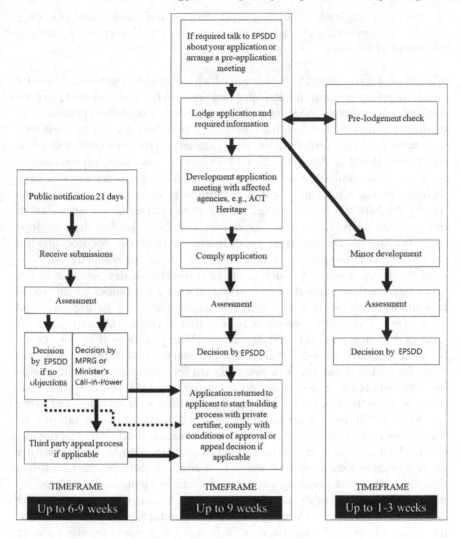

Figure 5.1 Development application process.

Sources: Complied from EPSDD DA process (2020)

He added to justify his comments on Technical Officers:

> The way they [technical officers] talk and the take the developer's part during any brainstorming session indicates that they are close allies and have common goals in development. We are just trying to do something for the betterment of the community. It indicates that our advice is not taken into consideration; rather planner takes advice from the proponents to initiate development work. That is the reality of Canberra's

planning situation. Planning Authority is not doing any planning; they are approving the developer's proposal only. Everything is running on an *ad hoc* basis.

Therefore, it is important to determine which stakeholders within the broader planning policy community actively participate in the decision-making process and who attempts to influence the overall decision-making process from behind (Dickinson, 1999). Some Committee members (except business representatives) and the wider public agree that development proponents such as the Master Builders Associations, GH Shaw and Associates, representatives from real estate and other architecture and planning firms are automatically included on the Committee, either as business representatives or as appointees by the Minister for Planning. These representatives have a great deal of influence on decision-making. Committee members also feel that development proponents are the main policy participants in the decision-making process, because proponents have vested interests in the outcomes of planning decisions, outcomes which could be a multi-unit development in the inner city area or approval of dual occupancy on a big block, which would benefit the proponents who construct houses and units in high-demand areas. Thus, Committee members believe that Technical Officers and high level of planning officers in the decision-making process within the administration regularly consult development proponents before DAs and Master Plans are tabled to the PACs or public forum for consideration. Some members commented that developers do not bring any DAs without prior consultation with Technical Officers, and the planning authority does not bring any Master Plans (which are normally designed by the planning staff, not by the development proponents) without consulting and obtaining great support from at least some developers. In fact, developers are essential in the planning process because their support is needed for any new policies affecting urban development. If the planning authority tried to implement planning policy without the support of the developers, this may result in a time-consuming and complex process for planning authority. Therefore, resident and community groups expressed concerns that Technical Officers and developers were the most influential groups in planning decisions. However, a business representative interviewed felt that resident and community representatives were not sufficiently competent to effectively influence the Technical Officers in consideration of their concerns about the planning decisions.

Another business representative, also a developer, commented that residents and community groups had difficulty articulating their values using appropriate planning language and terms. They only expressed emotional attachment to their places on redevelopment issues, which they think may affect them directly. In such cases, only vocal members on the committee may take advantage of this situation. A few members normally discuss the complete agenda and topics tabled for discussion and give no chance to others for equally participating in the discussion. These vocal members were drawing attention away from important planning issues or as a form of opposition to planning proposals. All the silent members are marginalized in the discussion, even though they are often asked to say a few words at the end of the discussion. This situation has been seen by a former Committee member as caused by their lack of ability to strongly and articulately express concerns over the discussion table. A member of Majura Committee commented:

> Convenors, planners and developers talk too much on issues and spend most of the time allocated for the discussions. When the time is finished, they look at us for our comments as if there were two groups—vocal and silent.

Some members commented on the influence of other organizations on the planning decisions, and identified residents and community associations within the local area planning regions as lacking the political power to have important influence on planning decisions, because they have no entitlement to be consulted regularly, even though they are often consulted when the planning authority feels it appropriate. However, community organizations attempt to influence the planning process from outside the legitimate consultation body through attending meetings, writing letters to the media, holding media events and inviting the planning authority to present any urban revitalization programs. This clearly indicates that community groups have less impact on planning decisions than do the Committees and the developers. The perceptions of Committee members on influence in planning decisions and the actors in the planning process are different. Interviews with a Committee convenor (also a community leader) indicate that all the planning stakeholders and other development proponents have some degrees of influence on the planning process; however, the community at large feels that developers enjoy the most.

The wider public consists of enthusiastic people who normally belong to various residents' and community associations. They often come to meetings and participate in the discussions, as they have been encouraged to do. Even though some groups have formal representations on all six Planning Committees, other members interested in planning also attend meetings on various occasions; some go regularly. But in a separate way, these groups have no formal power to influence decisions and they believe that planning authority has placed them far from the process of decision-making. Most of the Committee members (except the business representatives) and members of the resident and community groups feel that they are marginalized in total planning decisions. The Technical Officers recognize that these groups can register their concerns on planning and development matters through their representatives on the Committees, but they come to the meetings to support their representatives. This process delays the approval of development applications and creates misunderstanding among the policy community.

To evaluate the level of influence by the planning policy community on planning decisions, it is important to identify the role of formal consultation bodies and their places in the decision tree (Figures 5.2–5.4). It is also important to

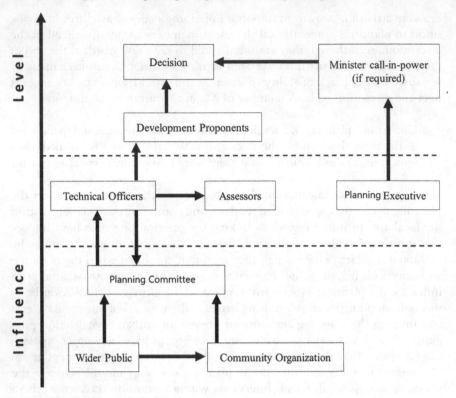

Figure 5.2 Influence on planning decisions (identified by Committee).

determine the status of groups in the consultation process. In the case of Planning Committees, most members do not agree on where the Committee stands in the decision-making process, compared to others in the planning policy community. There were many views of Committee's place on the decision tree that placed Planning Committee in different locations within the broader decision-making process. A member felt that advisory committee was naturally included in the planning process, and thus was a part of the decision-making process, and those Committees helped to bring community concerns using this platform into the public sphere. Two others members pointed to the role of Committee as facilitator or mediator between the community and developers, placing it just a step below the parallel to the Technical Officers and assessors (Figure 5.2).

However, some of the Technical Officers interviewed feel that Committees were created with the aspiration that various planning and development issues will be discussed before final policies and decisions are made. They also note that Committee provides an opportunity to register their concerns on planning and development matters and thus is situated on the boundary between development proponents and the wider public (Figure 5.3). All Technical Officers see Committee's main role as to make comments on development

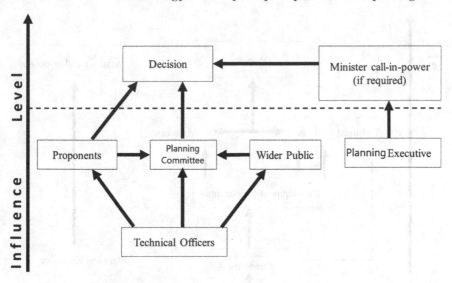

Figure 5.3 Influence on planning decisions (identified by Technical Officers).

applications referred to them by the developers or planning authority, but not to generate any new planning issues for consideration. This view reveals that Technical Officers place Committee on the outer periphery of the Planning Authority's planning processes.

All the remaining members of the Committee believe that members have a very limited influence on the final decision-making process and locate Committee at the bottom level of the decision tree (Figure 5.4). They feel that Committees have provided a great opportunity to the residents, community groups, and the general public to appear before the meetings to discuss and raise concerns about planning process, but they have no meaningful influence on the decision-making process.

Therefore, it seems that the fairness of Committee must be evaluated on two features. First, whether the discussion process is a good platform for public participation, where the community, affected people and the wider public are given adequate opportunity to raise their concerns and feel empowered through the consultation process. Second, Planning Committee as a community input mechanism which can assess the level of influence it has as perceived by the community and development proponents.

In providing an opportunity for members and the wider public to join the discussion process at the Committee meetings, Committee has been very successful as a platform of public participation. In the Committee meetings, the wider public can participate in discussion on planning and development matters. However, they do not bring any planning matters for discussion; they only discuss planning applications that have been conceptualized either by planning authority or by developers. They

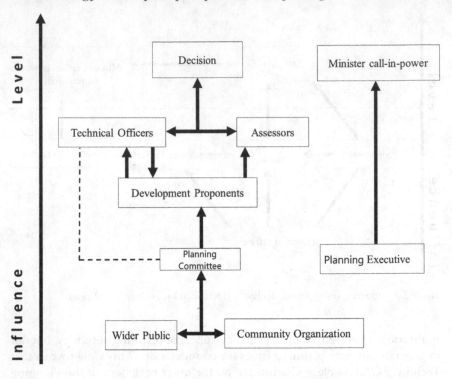

Figure 5.4 Influence on planning decisions (identified by wider public).

acknowledge that this gives them a platform from which to express community concerns, which they might not otherwise have had. Some members of the wider public interviewed believe that Committee provides the planning authority and the ACT Government with a level of legitimacy that the authority could not obtain regularly through other consultation mechanisms. In fact, some placed Committee much closer to the planning authority, as an equal partner in planning decisions than did the members of Committee. The wider public perceive that the planning committee is a good opportunity for public participation and that they have some ability to influence final decisions. Other consultation activities such as information nights, public hearings, submissions and public workshops, while less expensive, do not achieve the same level of satisfaction among the wider public.

Opportunity to participate in the discussion process

Webler (1995) notes that every participant should have an equal opportunity to attend, initiate and discuss in the consultation process. The interviews with members reveal that all members and the wider public have an adequate opportunity to express concerns on development proposals. However, some long-time members have been under prolonged stress and anxiety because they felt that the planning authority paid hardly any attention to their comments and preferences. They were given the opportunity to make comments on development applications but did not know whether the community views were taken into consideration. They have the impression that the preferences designed by the architects and planners would be approved with some expected changes, identified by a Committee convenor as 'minor changes'. They also have the perception that it is all pre-determined by the developers to give an impression that comments and community concerns are taken into account. He added:

> Changes are pre-planned, that will give an impression to the community... [that] they have been respected.

The interviews with residents also show that what matters to them is not the opportunity to express their views, but rather the importance of being listened to, and having their concerns taken into consideration. There was a feeling in some long-time members, particularly resident representatives, that the consultation was just an 'obligatory' process to give an impression that community consultations were carried out and decisions were being made with community input. They believe that the initial decision has already been taken without proper recognition of the community concerns.

Another community representative agreed that they are only given opportunity to participate in discussions at meetings or other consultations, but they are never invited or included during the conceptualization of a development application. Rather, planning issues were conceptualized either by the planning staff or by the development proponents, not by the community groups or other consultative bodies. Resident and community groups feel that their concerns are neglected, and developer's views given importance into the planning decisions. Some note that the planning authority and Technical Staff eventually take the community views into account for planning decisions, but developers do not adequately act upon the Committee's concerns and give them any real weight. A member commented:

> The planning decision-makers have the political stand and are motivated by the development proponents with an interest to serve them (the development proponents), not the community and residents.

A developer also commented on the issue and supported multi-unit development in the inner-city area, which he believes has a high demand for affordable housing. He added:

> We are responding to the market demand for affordable housing units, which will definitely drive the urban economy into a good shape. Young people don't want to go further to the Gungahlin area. They want to live

in close proximity to their workplaces. The Territory Plan also allows us to put dual occupancy in a big block and to put some more affordable units.

5.2.2 Evaluating fairness: Early involvement

The LAPAC members interviewed indicated that they would prefer a broader role that would engage them at the beginning of the planning process to register community views and act on behalf of community. But the existing process, they believe, only consults them at the end of the decision-making process, thus preventing them from making meaningful comments on development applications. An interview with a member indicates that Committee has experienced a great deal of frustration from the virtual impossibility of implementing some of its recommendations, as the Committee is involved too late in the consultation process.

However, some members do not want to be involved in the early phase of planning, which would require them to attend inter-departmental planning meetings to raise community concerns. They are overwhelmed with many DAs for the scheduled meetings and it is a voluntary task for the betterment of the community. Instead, they want various departments to come along periodically to the meetings for a presentation of future perspectives and development work in their areas of concern. A Committee convenor commented:

> I have got no time to attend various departmental meetings. I am overwhelmed with many development applications.

In contrast, some members want to be involved in the planning process by organizations such as the Chief Minister's Department, ActewAGL, Urban Services and other development proponents. A member said that the committee should be involved early in the planning process with various departmental planning meetings to provide community information. In this way, planners would be aware of the perspectives and attitudes of general community and allow more flexibility for necessary changes. A Technical Officer also realized the benefit of Committee members as well as others in the planning policy community being involved at an early stage of the planning process, because not involving Committee members earlier may result in the apprehension that their contribution will not influence the DAs. An ideal situation envisages the involvement of planning stakeholders before the development of the DAs in order to assure the Committee members that a fair and early consultation process is occurring. A Committee member said that this was not the case in the present type of consultation; rather the planning authority came along with the proposal and said 'that is what we have decided to do. You should have it'.

A Technical Officer gave his opinion as well: 'that the community should be involved at the early stage of planning, and planning authority is trying to bring in the community in the first place for the planning decisions'.

However, he noted that Canberrans were less interested in participating early in the process and discussing planning matters, but the wider community raised an outcry when they found that proposed development applications and other urban revitalization programs would affect them.

Interviews with residents and community representatives indicate that they want to be involved in the early stage of planning initiatives and meetings with various departments. On the other hand, most of the business representatives interviewed said they did not have enough time to attend inter-departmental planning meetings. Instead, they wanted departmental organizations to attend meetings for an informal discussion on possible development applications that would significantly affect neighborhoods. A Ginninderra Committee member commented:

> Well, Lawson [a suburb, yet to be released during the interview] will be released for residential development. We have only heard from people and seen if in the paper, and later developers come along with a plan. It seems they have done so much background work before coming to the community for consultation. They should have come early for a discussion with us.

However, as mentioned above, a Technical Officer commented that only a few Committee members, wanted to be involved early in the consultation process and that the community at large was not very interested in participating in the government's various consultation processes. The comment of a Technical Officer is pertinent:

> Canberrans are lazy. They do not come early in the consultation process. They come when they feel affected. We try to involve them as early as possible, but they do not turn up until something has happened in their backyard.

Most of the resident and community representatives have commented that early involvement of residents would give the planning authority a greater opportunity to incorporate local knowledge into the planning proposals. They feel that inclusion of local concerns into the planning decisions would ease the conflict between the planning authority and the community. A Ginninderra Committee member recognized the importance of local knowledge being accommodated in the planning decisions, and added:

> The population of our catchments [Macquarie, Aranda, and Cook] are ageing. We know how our local centers and environments work. But the proposed redevelopment of the Jamo [Jamison Group Centre] has been carried out by the planning team who hardly come to the centre. It is difficult for an outsider to figure out our needs such as needs for the ageing population.

Other members also acknowledge that they have the better option for the Jamison Group Centre redevelopment. Nevertheless, the planning team proposed three options, asking them to comment on each one. The community believes that this is not a fair process for consultation, and that they should be consulted before the formulation of the three options for redevelopment. A resident member of Ginninderra Committee who was also an architect-planner commented that the planning authority could have organized a charrette design workshop before conceptualizing the group center redevelopment plan and that the community could in effect design its own options for the redevelopment. Instead of approaching the community, the planning authority designed three alternative options in consultation with developers, and now the community is asked to choose one of those options. All options have multiple features to accommodate many units around the group center to meet the current housing demand in the area, but people in the community are dubious about putting many units around the group center, which they think is not the fair solution for the need to revive declining group centers. They believe that the community and residents should have been consulted earlier to produce various options. Some members commented that planning authority appeared to give importance only to the development proponents' consideration to meet so-called market demand and to politics above the local values for the planning decisions.

Similarly, an Inner North Committee member interviewed feels that the decision-makers do not consider local concerns early in the consultation process, because they are not considered professional, not articulated from the planner's table, not designed from the architect's drawing board, but considered as emotional expressions about places. The community is not involved because the planners consider this unnecessary, as reflected in a comment made by a Technical Officer:

> The residents' view is about the emotional expression of a place. Sometimes it has facts, sometimes it does not. They would be happy to see various options for them to choose if they can find something interesting, which we can consider valuable to consider.

It is rather important to identify the levels of involvement in the consultation process. Though there was a formal process for approving DAs, most residents and community representatives believed that planning authority had already made decisions by the time DAs were presented to Committees and other community groups for their comments. In the mandatory procedure for consultation, residents are officially given the opportunity to comment on predetermined and decided development applications. The community is not involved in the early process to determine the nature of the proposed development applications rather they are given the opportunity to choose among various options which are developed and conceptualized by the planning staff or development proponents, not by the community. A Technical Officer commented:

All Planning Advisory Committees should be involved in the consultation process when a complete set of Master Plans or DA proposals is being prepared for public comments.

Most of the planners believe that if Advisory Committees gets involved at the early or preliminary stage of planning, it will be difficult for the planning authority to complete all development works within a financial year. The community concern is to be involved with the planning process at the beginning of the formulation of development applications, which will provide opportunities for necessary amendments before the formal consultation process. However, the current process is identified by the Committee members as the one where they are involved at the end when major decisions have been made. Only minor changes are being endorsed to give impression that community concerns are being considered.

5.2.3 Evaluating fairness: Representation

One of the important criteria of fairness in public participation is the nature of the people involved in the process, so it is pertinent to analyze how people are recruited on the committee. What selection criteria are used? Do they represent cross-sections of the population? Do they obtain the wider community's views to put into the decision-making process? Who has a chance to take part? Who raises their voice? Who keeps silent or is excluded from the process? This study has examined these questions to evaluate the fairness of representation in PACs. One of the most often cited criticisms of the advisory committee is its lack of representativeness, because the Advisory Committee is perceived to offer the opinions of only a few people selected, elected or appointed by the authority purposefully from the resident and community groups.

Balance of membership on Planning Advisory Committee

Committee membership has evolved considerably over the years. Since its introduction, the ACT Government has changed the system of representation on all PACs. Originally there were two representatives from each suburb, three business representatives from all the suburbs, and one representative from the community council. Later, the ACT Government changed the nature of representation, adding one business representative from each suburb and continuing with Ministerial appointees up to three or four in one Committee.

Committee members are elected for two years with representation drawn from elected resident representatives, the local resident groups, the community council and the business community. It was found mixed feelings among Committee members about the current balance of representation on the Committee. Most of the resident representatives feel that current membership is not balanced. Some business representatives and Ministerial appointees normally do not live in the area, but only become members of Committee to serve their business and construction interests. However, the interviews with business representatives show that they are satisfied with the current balance of membership and its selection process. The membership of the Committee includes a wide variety of local environmental and planning activists, educators and businesspeople. Vari (1995) found that Advisory Committees with a broad composition, including professional disciplines, non-government organizations, academics and others, improved the Committee's level of impartiality and credibility while fostering a consensus-seeking process. In the case of PAC, this is true to some degree, as the interviews indicate that the Committee consists of various people involved in planning and development matters and other interested members of the public. However, many of the Committee members who belong to the resident and community groups are dissatisfied with the current selection policy, process and the balance of representation.

Committee members, who expressed dissatisfaction over the membership and selection policy, identified various reasons. They are very critical of the Ministerial appointees and business representation, with whom they do not have amicable relations. A former Committee convenor was critical of the vacancy filling system:

> The Minister normally nominates people like real estate agents and builders as residents. If there is a vacancy, he nominates landscape architects in private practice. Also, he nominates people from construction firms who neither live nor have an office in the area.

Another member raised the question of too many architects and planning professionals on the Committee, who served only the planning authority and overlooked community concerns. He added:

> Developers, people in the architectural industry could in effect dominate the membership of the Committees, even though the meeting is open for all and a lot of the community does not know, but developers and architects know it could be stacked in their favour. However, the process is fair. They do advertise the meeting in the papers. But there is an inability for a particular group to stack the meeting.

However, another Committee member was very positive about the Ministerial appointees on the Committee:

> The Minister can appoint people with knowledge of planning and design standards. Appointees are not in a dominant position in LAPAC anyway. He [Minister] can appoint people who probably would not go through the election process. Because they are not known but want to make a contribution and they are not dominant, so in that respect, it is all right.

Vacancies are another concern to operate its consultation process effectively in a timely manner. There are always large numbers of vacancies in all Committees: some have a continuing problem of quorums to conduct regular meetings, due to the regular absence of business representatives and the Ministerial appointees from monthly scheduled meetings. It was noticed that the Majura Committee meeting was cancelled several times owing to the absence of Committee members, mostly the business and the Ministerial appointees. In this regard, a Majura Committee member commented:

> Businesspeople attend the meeting if the agenda has interest for them. Otherwise they send an apology to the convenor. ...If the committee is operating in a vacuum, then there is little scope for improvement of outputs and committee process.

Therefore, Committees lacked consistent contributions from all sections of the community and lost potential publicity about its existence within the neighborhood and greater community.

Selection criteria

There are no formal selection criteria for the Committee members to be included on the committee except for verifying the ACT driver's licenses of the interested people entering the meeting room on election night in order to ensure the participants' resident status of a suburb covered by a Committee. This process is fair and democratic, since everyone has an equal chance to nominate anyone to be a member of the Committee. However, there is no clear statement of selection criteria except for residence in a neighborhood and business interests in those neighborhoods covered by a Committee. It was observed attending every Committee election for three years that planning authority did not have any selection criteria to target credible and knowledgeable individuals with knowledge of planning and design to be members of community-based PACs. On the other hand, there is an Advisory Committee in the ACT Government, named the Environmental Advisory Committee (EAC), which is based on a set of selection criteria (Environment ACT, 1999). The EAC is an Advisory Committee comprising people with experience in the environmental field, who can give informed comments to Environment ACT for consideration. This is not the case in PACs.

Interviews with the planners indicate the importance of members from all parts of the community in Committees and the importance of accommodating diverse views on planning matters. Planners also recognize that Committee members have been giving free consultations to the planning authority on a monthly basis, since Committee members are not given any honorarium for their involvement. They spend time and effort in reviewing and providing advice on development applications and fulfill their roles very well. Accordingly, the planners are satisfied with the level of detail in Committee work. However, they believe that it is not the responsibility of the community to nominate people with expertise to PACs, as this would increase the complexity of the planning authority accommodation of all comments and concerns.

The interview results show that the *election process* is fair and democratic. Every member of the community has the right to participate in the election to be a member of a Committee and nominate anyone to be on the Committee. The Minister has the discretionary call-in-power to appoint anyone that the Minister feels appropriate for the s. The provision of appointees has created doubt and anger among the elected members, particularly among the resident and community groups. Some of them raised question of their importance in the community consultation process:

> The election process is quite democratic. However, the appointment process is not good enough. I don't understand why Mr. Roberts [not actual name] becomes a member of this LAPAC. He neither lives nor has any businesses in our LAPAC areas. He has a business outside our area and how he became a member of our LAPAC, he said, he was appointed by the Minister. I should have told him you are disqualified. But I know him; for this reason, I cannot speak to him like this. Even if you have a house or a shop [in LAPAC area] and don't live here [within a LAPAC boundary], you could be a member of LAPAC.

Another member of the same Committee expressed similar view:

> The process [of election] is democratic; when it comes to loading the committee with businesspeople and Ministerial appointees, it is not correct. People living in Red Hill that have a shop in Dickson, are eligible to be a member of a Committee. I feel it is not correct. We want businesspeople, not a shop-owner who neither lives in this area nor has familiarity with the neighbourhood.

On the other hand, another member acknowledged that the current election process is fair in terms of the number of people attending the election, but raised question of the low profile of the election process in the community; she suggested that the current process requires structural changes to ensure the presence of more members of the community at the election. She added:

> The election procedure needs to be very democratic, for instance, there were two votes only for a candidate, and the supporter was the wife of the candidate. I believe this is not a democratic process. There should have a system to ensure greater presence of residents, community groups, school representatives, church leaders and businesspeople during the election process, which is totally absent in the current LAPAC process.

Sometimes a business representative does not truly represent the area of concern where the business has no relevance, such as in the Parliamentary Triangle and Barton, where there are many government offices; however, there is a business representative for Barton. The residents and community representatives believe that this type of membership normally serves only the interest of developers and other development proponents, not the residents and the community at large, so most of the resident representatives do not want them on the Committees. They believe that business representation is necessary for a balanced membership of the committee, but they commented that there should be check imposed on membership eligibility. Some members expressed the view that Committees should only be a platform for resident and community representatives, not developers or businesspeople appointed by the Ministers. In this regard, a member commented:

> If there is a proposal to build a house in front of Federal Parliament, they [developers] will go there to build a house.

Some of the dissatisfied respondents were asked to identify the underlying causes of their dissatisfactions over the balance of the Committee structure. To explore their perceptions on membership and recruitment, all dissatisfied respondents were asked to explain their reasons. Most said that Committee had excessive Ministerial appointees and lacked representatives from all sections of the community and residents. They were concerned about too many business representatives being on the Committee. For example, although West Belconnen Committee had no appreciable business establishments, it included business representatives. They are also concerned about the non-representation of important stakeholders on the PAC. A Committee member pointed out the absence of NCA representation on the Committee, although the NCA is important in the planning decisions in ACT.

Nature of representation

Committee protocol provides for the inclusion of representatives from community councils, resident associations, and business groups. In addition, the Minister for Planning has the discretionary call-in-power to appoint whoever he or she believes to be appropriate for the Committee (PALM, 2000). Accordingly, every Committee has representatives from community councils, resident associations, community action groups and business groups. PALM invites community councils to nominate their own members to represent the council on the Committees. Normally, only one representative from each community council is officially eligible to represent the council, but sometimes more than one representative from the council attends the meetings. In Manuka Committee, it was observed that more than one member attended the meeting regularly, but other members were not concerned about it. Manuka Committee is the only Committee that allows multiple representations and its convenor explained this saying 'we are very much flexible'; however, the community council has only one vote. This flexibility has given the whole community a great opportunity to participate in the decision-making process.

The observation and interviews suggest that community councils provide stronger and more detailed explanations of current proposals than do other representatives at the meeting. The community groups bring with them detail proposals that will positively or negatively affect their neighborhoods. This is because community groups have regular meetings and access to information, and their members are experts from diverse fields including architecture and planning. They can generate a shared vision in planning for their neighborhoods in partnership with the planning authority.

In the Manuka Committee, there is a clear distinction between community groups and other representatives. The Old Narrabundah Community Action Group appears to have more credibility with professional architects and planners, and greater opportunities for input in the planning process than others in the Manuka Committee. The Old Narrabundah Community Action Group does regular research on planning and development activities, which may adversely affect residents. It often provides various alternative planning proposals if they seem relevant.

Similarly, in the Inner North Committee, the community groups and other interested community organizations have very well-defined views on planning matters. This Committee covers most of the important places in Civic and its adjacent neighborhoods, and its community representatives are actively involved in making comments on Inner City revitalization programs. Members of the community council regularly attend Committees meetings with other organizations that have interests in planning and development. Planning for ACT Together (PACTT) is one such planning organization, whose main task is to deal with planning and development matters only. PACTT regularly organizes workshops and meetings on current planning issues, inviting a wide range of organizations that include political parties, developers' associations and academics in order to deliberate on planning and development. Views from cross-sections of the community are summarized and sent to the planning authority for consideration.

5.2.4 Evaluating fairness: Agenda setting and minutes taking

Webler (1995) identified agenda setting as a criterion to evaluate the fairness of the consultation process. In order to evaluate the process of agenda setting, it is important to evaluate its operational procedures for conducting meetings and discussion. More importantly, discussions are based on the available agendas. Members contribute throughout the discussion and make comments on individual items of the agenda. Vari (1995) notes that the agenda for the Advisory Committee activities is usually defined by the 'problem owners'. Many important issues may not be included in the discussion process. The 'problem owners' may select the agenda strategically, to support their own interests.

In general, the Committee coordinator, in consultation with its convenor, sets the meeting agenda. No scheduled meeting has ever been cancelled because of the lack of a meeting agenda, but sometimes meetings were cancelled owing to insufficient attendance for a quorum. This is because some business representatives do not attend the meeting regularly, as the meeting agenda may not interest them.

The coordinator includes on the agenda all recurring issues – any issues referred to Committee through communications from planning authority, requests and correspondence from any outside group or individual, and any item added by the LAPAC convenor or other members. All members of the Committee interviewed feel that they can add items to the agenda at any time and feel comfortable with the agenda setting procedure. But some of the members are concerned about its appropriateness. They feel that the convenor, not the coordinator, should have the final say in adding any item to the agenda for discussion in the meetings. Initially, the Committee used to prepare the agenda for discussion; later planning authority took over this responsibility. An experienced member of a committee expressed doubt about giving sole responsibility to a convenor for final agenda setting. He added:

> The previous convenor of this LAPAC was rigid to include any item during the meeting, even though there were issues of current importance to discuss. Although he was a residential representative, I got the feeling that he was actually a lobbyist for the developers. He had an office in the developer's office. Surprisingly, he was using that office's stationery and even email facilities. So, sometimes they acted on the developer's behalf.

On the other hand, a Burley Griffin Committee member believes that the convenor has a very small role in agenda setting. It is the Committee coordinator, who, along with developers, normally lodges development applications and sets the agenda for discussion in the meeting. Because the planning authority and development proponents have planning issues to be discussed, the community has nothing to put on the agenda:

> The community has no role in putting the items on the agenda. LAPAC is reactionary to put MP [Master Plan] and DA. We are simply reacting on the MP and DA. For this reason, we are reactionary.

Time allocation for agenda setting

Some members are concerned about the time allocated to the committee members for the agenda. Sometimes, the agenda reached the members on

Thursday or Friday before the scheduled meeting on Monday. Sometimes, it did not arrive until Monday morning and that would be the first time the members of Committee knew something was on the agenda. On a few occasions, the planning authority delivered the agenda in the afternoon and most of the Committee members could not manage time to look at all development applications. For these reasons, some members who are normally very regular in attending the meetings feel frustration about the time of agenda setting and its delivery. They believe that this is intentionally designed not to give advance notice to the members, in order to avoid informed suggestions that will probably conflict with pre-conceived ideas, or to avoid people's concerns. A member raised this issue directly to the Minister for Planning and Urban Services in a public meeting and asked him whether it would possible to make the agenda available, perhaps a week in advance. In reply to her question, the Minister said, 'That's a good suggestion, and we will take a look at it' (PACTT, 2001).

Agenda format and process

Before the meetings start, there is an opportunity for every member to announce whether they have anything to put onto the agenda. The first item on every meeting agenda is 'welcome and apologies'. The Committee coordinator records the presence and absence of current members at the meeting and introduces any delegations and parties that will present and clarify development applications and Master Plans. A wide variety of groups and individuals appear before the Committee occasionally, including citizens or groups with specific planning and environmental concerns. The Committee coordinator is always accompanied by a Technical Officer, who normally discusses and explains the PALM position on existing DAs or Master Plan, and explains technical issues where necessary.

The second item on the agenda is 'Declaration of Conflicts of Interest', followed by the approval of the minutes from the last meeting of the Committee. The third item is 'LAPAC Correspondence', in which the committee considers any correspondence it has received before moving on to consideration of new and old business. The Committee regularly receives correspondence from organizations and agencies in the Committee areas, including planning and environment related organizations such as Save the Ridge, Save our Suburbs, Hackett Interim Group, Griffith and Narrabundah Community Action Group, The Belconnen Community Council, Resident Associations and Planning for ACT Together (PACTT). This correspondence keeps the Committee members informed about planning and environmental issues and activities in the Committee regions. The fact that Committee receives and reviews correspondence from a broad variety of local groups and businesses indicates that Committee is very well informed about planning issues in Canberra and monitors the progress of many urban policies and programs.

The fourth item is 'progress report on outstanding DAs'. Sometimes committee members ask the developers to provide necessary information regarding the proposed development applications. The developers normally attend with the requested information, which, in some cases, the opinions of the people possibly affected by the proposed development applications. This item is followed by the presentation of a new development application by the concerned developer.

After all the business items on the agenda are discussed and acted upon, the floor is open for consideration of any other matters tabled by the coordinator or members of the Committee. At the end of the meeting, the convenor reminds all members of the date of the next meeting.

Agenda setting: Too much work

All six convenors along with two previous convenors interviewed were happy to be a part of Committee and felt that they contributed significantly to the planning decisions on urban revitalization programs. Conveners also believe that through the consultation process, community input is generally taken for the planning process and considered as community views. They acknowledge that they have a social responsibility to act on what best suits the whole community. However, they feel overwhelmed by looking at many development applications, correcting previous meeting minutes, communicating with other members and preparing their own concerns about all agenda items that are to be discussed in the following meeting. Some members feel that it is too much unpaid work. Also, the area of concern for a Committee area was so wide that sometimes members spent much of their personal time in looking at all development applications. A planning spokesperson of a political party commented:

> ...it's fair to say that LAPACs already have too much work that they can
> handle and I note that this is more a function of the Government's refusal
> to properly resource LAPACs to undertake their terms of reference, that
> it is of the capacity of people on the LAPAC to do that job well.

Agenda setting: Hidden agenda

Some resident members interviewed feel that business representatives have hidden agendas to be approved by the planning authority. This is evident in the following comment made by a Burley Griffin Committee member:

> The community sometimes became disillusioned when their comments
> were not taken into consideration due to some hidden agenda. They
> [developers] were not transparent. All appointees have got their own
> hidden agendas to support developers and proponents or Authority's
> planning initiatives. They [planners] are not transparent either.

Dual occupancy development involves the placing of two dwellings on one big block. Currently, there are many applications for dual occupancy development in most of the suburbs close to the inner city area, but Committees are not officially allowed to make comments on dual occupancies or to some extent on heritage issues, which are very important to discuss in the meetings. However, Committees are not entitled to make comments and advise on dual occupancy, transportation planning, land lease, land sale and release and heritage issues, which planning authority believes should not be referred to the Committee for consultation. An Inner North Committee member made a comment that reflects total dissatisfaction over the exclusion of planning matters:

> Dual occupancy is in our view, the single biggest development issue in the Inner North. Yet the LAPAC is deliberately excluded from considering dual occupancy proposals and, therefore, from informing the Minister/PALM of its and the wider community's views. This severe limitation of the LAPAC's role as a community consultative body is all the more significant in the light of PALM's apparent policy of judging each application on its own merits, with no consideration of the cumulative impact of the ever-increasing number of dual occupancies.

Another Committee member commented:

> We are consulted on various issues such as fence height, extension of garages, and extension to put additional rooms on existing house, however, the funny thing is that we cannot make comment on dual occupancy—that is dreadful. It seems dual occupancy is not a planning issue.

Dual occupancies were certainly an issue when the Committees were established in 1995. But the government specifically excluded dual occupancy, as there was a concern that it would have generated too much work for the Committees and planning support staff. In practice, individual members of Committees could provide comments to the planning authority on specific dual occupancy proposals, but the Committee as a separate entity could not. Consultation protocol specifically mentions that members are not entitled to make comments on dual occupancy. But members feel that dual occupancy developments have a considerable impact on the amenity and character of local neighborhoods. For these reasons, a planning spokesperson of a political party raised this issue in the ACT Legislative Assembly:

> Mr. Speaker, …[party] thinks it's time that LAPACs, as representatives of communities affected by redevelopment activity, should have the opportunity to comment on dual occupancy development and to start giving PALM and the Minister some feedback on exactly what their concerns are about the nature of dual occupancy.
>
> (ACT Legislative Assembly, 2000)

Despite many requests to include dual occupancy in the meeting agenda, the industry groups do not want to refer it to the Committees. A member of the wider public, working in an architectural firm, sent the following letter to the Minister for Planning which was read out in the ACT Legislative Assembly to justify the non-referral of dual occupancies to the Committee meetings:

> I recently attended my first LAPAC meeting. I was appalled by the negative, anti-development attitude of most of the committee members. I refused to believe their views are representative of the broader community. As a member of the Manuka community, I must applaud your position on the non-referral of dual occupancy proposals to LAPAC.
>
> (ACT Legislative Assembly, 2000)

When this letter was read out, most of the Members of the Legislative Assembly made comments on Advisory Committees and its involvement in planning decisions. However, a convenor of a Committee expressed anger and dismay over the issue and commented:

> The Minister read out a letter, which went against the LAPAC process and the attitude of LAPAC members, but the Minister never read out any letters that we sent him stating our concerns on planning decisions.

This convenor also believed that the Minister was against her Committee, which was clearly demonstrated in the ACT Legislative Assembly. She also commented that enjoying a planning portfolio in the ACT Government did not reflect commitment towards the community inputs and concerns. A resigned convener made a similar comment concerning the Minister's action in the ACT Legislative Assembly:

> I have sent him [Minister for Planning] letters concerning imbalanced selection criteria of the member and other important issues, which have never been referred to the Assembly. Only negative propaganda has been given high priority and importance by the Planning Minister on the planning situation in ACT.

Minutes taking

Writing up conversation involves a lot of work and mistakes are inevitable. Minute taking is a very important aspect in the consultation process. It is treated as a community input for the planning authority and for the Minister for Planning in order to incorporate community views in the planning decisions. And often, the planning authority depends on minutes to identify community concerns and to understand community preferences on a development application. Furthermore, minutes can be treated as the total advice on the issue to be considered by the planning authority as true community reflections. The concern is that there is only one coordinator for all six Committees, who normally records all minutes.

After recording the minutes, the coordinator normally sends these to the respective convenors for confirmation. Some members value the minute taking, as they believe it has importance in the decision-making process. A Committee member made the following comment:

> I have a major difficulty with the quality of minute-taking at LAPAC. I believe that often the importance of certain decisions/agreements made by the LAPAC groups is not accurately carried over to the minutes. A lack of focus by the group at times could be blamed for this. Sometimes, it would be difficult for the minute-taker to recognize that a decision has been made.

The main concern about the minute taking is accuracy in recording facts and opinions given by the members. Minutes are also valuable to the assessors when examining the merits and demerits of development applications and to incorporate community concerns. Therefore, the accuracy of the meeting minutes is an important issue in the consultation process. Since Committee is a group of diverse people, they certainly express diverse views on a specific topic. It is difficult for a coordinator to accurately record all comments made by the participants in the meetings. Sometimes, the coordinator gets confused on whether the Committee has reached a decision, and thus the coordinator often makes mistakes in recording comments. A coordinator commented:

> I always get confused about whether there was a decision. I therefore record all comments onto the minutes.

A member reported that sometimes the coordinator wrongly records comments on certain issues. Most of the time the minutes need editing at the beginning of the following meeting in order to record actual comments. In this situation, members who are absent from the next meeting have no chance to make the necessary corrections or to check whether their views are adequately recorded. The conveners, who are supposed to look at the minutes before going to the planning authority, are believed to be overwhelmed with development applications and the necessary preparations for the next meeting. There is no checkpoint to recording the minutes whether they have been recorded accurately and neatly. Committee members often need to correct reports of their own views in the last meeting on issue-specific discussions.

Role of the moderator

The Moderator and the facilitator in any public consultation process play an important role. Both serve in the consultation process as neutral entities responsible for enforcing the rules as indicated in the consultation protocol. There is, however, a difference between moderators and facilitators. A facilitator merely tries to discuss the specific agenda without guiding it, while a moderator demonstrates more leadership. A facilitator may encourage silent participants to discuss issues, while a moderator may make proposals, participate in the debate by presenting relevant information on the current agenda, and present arguments on many elements, which are missing from the discussion (Webler, 1995). The attitude and behavior of moderator or facilitator should be subject to the scrutiny and approval of the participants in the consultation. Webler's (1995) model of fairness outlines three ways to evaluate the role of moderator and rule enforcement. For him, everyone must have an equal chance to suggest a moderator and method for facilitation, challenge and support suggestions by others for a moderator and a facilitation method, and influence the final moderator and facilitation method selection.

In the case of Committee, it is neither a moderator nor a facilitator but a convenor who conducts the meetings. The convenor is also one of the parties affected by the development applications, but the moderator or facilitator may not be a member of the affected parties. The Committee convenor is involved in the Committee in the capacity of a resident, community group member or business representative. With regards to Webler's (1995) fairness analysis model, it is not possible to fit all the categories into the Committee process. It is implied that the Committee process is fair enough to suggest a convenor for a Committee; however, there is a question of the method for facilitation. In such situations, the Committee process should be evaluated with the role that the convenor normally plays in conducting meetings, and organizing all information relating to the meeting and the discussion process.

Conducting meetings and committee operations

The convenor chairs the meeting. In the absence of a convenor, this is normally done by an experienced member. The convenor is elected and reconfirmed by the Committee at the beginning of a new term, with a maximum two-year term for each convenor. The convenor has many roles in administering meetings, the main role being to facilitate meetings according to the established agenda and to maintain the Committee protocol. In general, the Committee meetings are run according to an agenda, although occasional items may be postponed to a later date if developers do not appear at the scheduled meetings to present and display plans on proposed development applications. In such case, chair of the meeting writes a letter or makes a resolution censuring the developer for not appearing. It was noticed this situation on a few occasions in Manuka, Inner North and Ginninderra Committees.

Many Committee members, who were interviewed, including the convenor and former convenors of six Committees, commented that it took some time for an Advisory Committee convenor to learn the regular operations of meetings, and communicating with planning staff and development proponents. Most of the respondents feel that having a convenor in place for a longer period is important in order to develop and maintain effective working relationships with the planning authority and development proponents, as well as to provide continuity in the operation of a Committee. However, some committee members commented that the convenor could be the only one for all Committees, but meetings could be chaired based on rotation. Others feel that rotation systems would jeopardize the direction and consistency of the meetings and the total process.

The interview data indicate that a clear majority of the committee members agreed that conveners manage the Committee meetings in a competent and professional way, but a few expressed dissatisfactions with the style of conducting meetings along with other activities performed by the convenor. Only the Inner North, Manuka and Burley Griffin Committees were somewhat critical of the convenor, while all other Committees were satisfied with their convenors and their overall performance. Most felt that conveners were not playing an important role and were not maintaining coordination with the planning agency staff to convey community concerns over the DA that would affect them so the original proposal could be reconsidered. Some were critical of the convenors and had the feeling that convenors were on the side of the development proponents and were not making efforts to protect their neighborhoods from bad developments.

Apart from the question asked about the overall competence of convenors, committee members were also asked to comment on to what extent they agreed or disagreed with the performance of convenors at various stages of meetings and its administration. These included statements about neat and succinct clarification of the agenda, giving ample time to members, flexibility to allow extra items on the agenda, and adequate understanding to explain DAs and Master Plans with facts and figures. Commenting on these objections, most of the Committee members acknowledge that meetings tend to be quite smoothly administered by the convenors except for a few instances in Manuka and Inner North Committees. What was revealed from the observations of these two Committees is a conflict of interest between business representatives and resident and community representatives. On the other hand, other Committee members who were interviewed agreed that the members of the committee work well together and are respectful of each other's comments. They said that most decisions were made by consensus and if there were no consensus opinions on any DAs, this could be recorded in the meeting minutes for the planning agency to consider as 'difference of opinion' on various development applications.

5.2.5 Evaluating fairness: Feedback and responsiveness

The PAC appears to be a good platform for public concerns and inputs, but planning authority is not equally responsive in considering and acting upon that public input. The end result is an Advisory Committee that generates the appearance of effective public participation. The perception of the wider community about the Committee process is that it is a positive addition to public participation. However, it lacks adequate responsiveness to the Committee's concerns. While the Committee functions well as a good method of public participation, it has only limited ability to act as a policy input mechanism providing advice to the planning authority and the Minister for Planning on many planning issues. The Committee is a way for the wider community to provide input to the planning authority, but its effectiveness is severely curtailed by the lack of fairness and openness in the whole planning process; because, although in principle the planning authority accepts the Committee advice, the majority of Committee members believe that Planning Authority fails to accord this advice any real weight. The Committee is established to deal with information provided by the Technical Officers and development proponents to planning authority or the assessing committee, not to filter public opinions. While policymakers of planning authority solicit public input from interested groups and individuals, they do not have any standard way of processing or responding to that input.

Feedback to committee members is an important factor to evaluate the fairness of the public participation process, but in the consultation process there is no structural form of feedback whether the advice is considered, modified or rejected. Even though the Committee protocol indicates that planning authority is to 'consider LAPAC comments on DAs and provide feedback' (PALM, 2000b: 3), it appears that planning authority seldom provides feedback or follow-up concerning consideration of Committee's contribution, or what decisions have been made. This lack of responsiveness between planning authority and Committee leads to assumptions that Committee views are not considered.

The interview data indicate that some members were critical of the process, as there was no formal way to provide feedback to the community and to the consultation Committees. Members felt that they did not get proper answers to their recommendations and also believed that listening to what people have said and then ignoring it could not be a good consultation process. This situation was described by a LAPAC convenor:

> There is no feedback at all from PALM, the Commissioner or the Minister. The attitude about the value of LAPAC input is not feedback. LAPAC members have no idea whether any of their comments and

submissions have affected any outcome. They have no idea why their comments and concerns are rejected. To simply say that LAPAC views have been considered is all but meaningless.

It is worth mentioning the words of former NCDC Commissioner at a public really:

> The community has no effective mechanism for having its views made known and being accorded due consideration by the Minister, as well as by the public service's Planning and Land Management agencies. While the Government and the Legislative Assembly pay lip service to public consultation, few changes are made as a result of LAPAC recommendations either. It is fundamentally a policy of deceitfulness that inevitably leads to frustration and disillusionment on the part of citizens and community groups who are accordingly becoming vocal.

Similarly, Committees do not monitor the outcomes of their recommendations; only enthusiastic individual members normally do some informal monitoring of development work. This individual questioning is not sufficiently regular to keep members up-to date on the current status of their recommendations.

The interviews with Committee members indicate that they are concerned about whether their comments and advice are taken into consideration. Monitoring is an integral part of any consultation process and keeps all concerned parties up to date on whether their views are being considered by the implementing authority. A systematic form of monitoring in the consultation process is essential, particularly where regular consultation is required (Rahnema, 1992). There is much criticism of planning authority that it has not been informing Committee members regularly about their comments and concerns, which leads Committee to the question of the viability of its protocol claiming their comments are taken into consideration. Providing feedback to the planning community is a statutory planning procedure that involves notification by letter to all concerned parties who comment on a development application or a Master Plan. However, interview data indicate that only a few receive response to their submissions and even then, they have to ask planning authority to respond. Even if planning authority provides feedback to the Committee members after repeated requests, this does not have enough information and contains no indication that Committee comments and concerns were valued and considered. In addition, feedback does not adequately explain why Committee concerns could not be incorporated into the planning process. Following are four extracts from meeting minutes, which demonstrate Committee concerns at not receiving feedback.

> LAPAC expressed concern that on a few occasions where feedback was provided to the developer, comments were disregarded, and concerns

were not addressed. Examples were provided including development on Lhotsky St. The LAPAC recommended access should be from Cartwright St to alleviate the high traffic generation from what is already a very busy road, however, this concern was not addressed.

(PALM, 2003c)

Improvement is required on feedback returned to LAPAC on comments made on development applications and we desire to see comments made by LAPAC to be heavily considered by PALM.

(PALM, 2003d)

LAPAC asked for feedback in relation to what extent were LAPAC comments taken for the decision-making process. PALM advised that whilst LAPACs have no right of appeal in terms of objections lodged against development applications, 'Notices of Decision' now address individual comments raised by members.

(PALM, 2003e)

...the discussions and material produced by the LAPAC do not receive adequate feedback and response from relevant governmental authorities.

(PALM, 2003e)

When asked by members for information during the meeting, the Technical Officers only orally reply to the members' queries without giving any firm decision. Even if they provide information, often it is not the specific information, which was sought by the Committee members. This situation is reflected in meeting minutes:

LAPAC members requested a list of DAs printed on the day of the LAPAC with a progress report included. Ray [a Technical Officer] advised the program is not set up to include a progress report with the list and can only continue to inform the LAPAC of progress verbally.

(PALM, 2003e)

All the Committee members interviewed were asked to respond to a question about what they would do if their recommendations and advice are modified or rejected. They normally protest planning authority's decisions in many community forums and write letters to the Executive Director or in some cases to the Minister for Planning. Some members accept the decisions because they lack the power to counter the decisions made by the Authority. In general, a majority of the Committee members expressed dissatisfaction over the monitoring systems and the feedback.

5.3 Chapter summary

The analysis of fairness in this chapter mainly focuses on the context and processes of public participation in PACs. The interview data reveals differences in opinions and ideas among the planning stakeholders in light of the specific criteria of fairness. In the context of PAC, the members and other involved in planning decisions expressed different opinions and ideas about fairness of public participation process. The planning policy community involved with the Committee process had various perspectives on the consultation process. The results in this chapter suggest that Committee concerns are not adequately taken into consideration, but this is not a universal opinion; business and Ministerial appointees have different views of specific issues of fairness process. Similarly, Technical Officers feel that they try to incorporate community input into the planning decisions but recognize other factors such as the high demand for housing in Canberra and the issues of urban consolidation. Overall, the community feels that they are less influential than other planning stakeholders in the planning decision-making.

The community representatives identified various drawbacks concerning the fairness criteria, of which, the unbalanced representation and the call-in-powers of the Minister are the main issues. They also blamed that the allocation of time was insufficient for the members to review development applications, and poor feedback policy, which made them feel alienated from the consultation process. Regarding the quality of the information, residents and community members commented that little effort was made in writing planning documents. Most of the members have the impression that planning documents such as Master Plans, Local Centre Redevelopment Plans, and Neighborhood Plans were written simply to present planning steps as required by the planning authority, but not with the intention of adjusting the information and writing style to the needs of the public. Because of this, members lacked the opportunity to participate meaningfully, which subsequently prevented them from learning more about planning, so they lost confidence in the consultation process. Thus, members suggest that adequate attention and priority should be given to using community language in planning and design documents. Extensive use of planning terms alienates them from making effective comments on planning issues.

The evaluation of the fairness criteria in the consultation process is summarized in Table 5.1. This book evaluated the fairness criteria on a three-point scale (SM, SS, and NS) indicating 'satisfy most criteria', 'satisfy some criteria' and 'not satisfy the criteria', respectively. This scale is also applied to the evaluation of effectiveness in public participation process.

Table 5.1 Summary of the fairness evaluation in the consultation process

Criterion	Evaluation	Brief description
Adequate Opportunity	SS	Members have opportunity to participate in the discussion but lack access to knowledge and information. Members also believe that they have little influence on the decision.
Representation	NS	Only selected and known people are involved. There is no uniform representation from all parts of the community such as church groups, the young, the disabled, knowledgeable persons, and others such as NCA representatives.
Early Involvement	NS	Committees are involved at the end, not at the initial stage of planning proposals. Planning authority organizes consultation meetings with other non-statutory groups and at the end planning authority comes to the Committee.
Agenda setting and minute-taking	SM	Committee members can put items in the agenda; however, some planning issues are excluded such as dual occupancies, transportation, land sale, and land release.
Feedback and Responsiveness	NS	There is no feedback system in the consultation process. Committee does not know whether comments have been taken into consideration and does not receive any explanation why their recommendations were rejected or modified.

Note

1. Equal access to information and equal access to knowledge are not similar. Webler (1995) differentiate these two aspects: access to information denotes the availability of information with facts and figures; access to knowledge indicates the adequate explanation of the data. For example, the level of PH value in water does not explain its effects on the environment: the data should explain how the environment is affected. This process is called 'access to knowledge'.

6 Evaluating effectiveness in public participation process in urban planning

6.1 Introduction

This chapter evaluates the criteria of effectiveness in the public participation process through a Planning Advisory Committee (PAC) by using interviews, observation and document analysis. All criteria discussed in this chapter were adapted from the available literature (Webler 1992, 1995; Landre and Knuth, 1993b; Beierle 1998; Octeau 1999; Renn et al., 1995; Lauber and Knuth, 1999; Rowe and Frewer, 2000; Webler and Tuler, 2000; Tuler and Webler, 2001) and from the observation of consultation processes in the PAC (hereafter Committee). This was established through attending Committee meetings over a period of three years, interviewing its members, and analyzing meeting agenda and minutes. The main criteria assessed are: the role of the Committee members, the decision-making process, conducting meetings, communication with the planning policy community, the effect of committee activities on the wider community, available opportunity for learning about the process, relationships with proponents and formulating advice to the concerned planning authority.

The aim of the criteria evaluation is to determine the factors of effectiveness in public participation in planning decisions. Such evaluation may offer several indications that can be adopted to guide future formulation of the participation process and may recommend guidelines for evaluating existing practices. The result of the evaluation may also provide feedback information to the planning agency to reveal the participants' perceptions of the strengths and weaknesses of the current processes. The perceptions of participants can also be used to formulate a new or modified process, while the weaknesses may show the way to make improvements in the participation process.

6.2 Criteria for evaluating effectiveness in public participation

6.2.1 Evaluating effectiveness: Defined role of participants

The PAC protocol states that the main role of the Committee members is to comment on broad planning directions for their communities. This can be

DOI: 10.4324/9781003122111-6

done through development of Community Value Statements, by commenting on development applications, as well as by other planning initiatives that may affect the planning and design character of their neighborhoods. More explicitly, the role of the members is to advise on planning and development matters, and to raise community concerns on the planning issues that help the planning authority to make informed decisions on urban revitalization programs (PALM, 2000b).

Current role: Committee members' perspective

All Committee members who were interviewed felt that the primary role of the Committee is to act in an advisory capacity in planning decision-making. The main task of the Committee is to provide advice and recommendations to the Minister for Planning on development applications and other planning proposals referred for consideration. The Committee holds no executive power to make decisions at the meetings and has no power to reject any development applications for further development; instead, Committee acts in an advisory capacity and makes recommendations on development applications. However, some members feel that they have a very distinct role in planning decisions as part of the community. In order to explore different views on the role of a Committee, Committee members were asked to evaluate their role in the planning decisions, and it was found that members were not in full agreement about their role in the process. However, most of them believe that the role of Committee members is advisory in nature. A few members think that they should be a watchdog in planning decisions. Some long-time members believe that the Committee is a fact-finding body to give community input to the planning authority, so it can make informed decisions on planning and development matters. It is widely believed that some members are appointed by the Minister for Planning because of their expertise in fields related to planning and design matters, and therefore, can provide a professional opinion on development applications; thus, the Committee acts like a fact-finding body for the planning agency. However, a few members believe that the role of the Committee is adversarial and confrontational. These are mostly the business representatives who have shown great frustration about the Committee activities and outcomes and believe that Committee's attitude in planning decisions is negative toward better planning outcomes.

One business representative expressed similar frustration with the fact that the position of Committee is low in the procedural process: He commented:

> If it is an advisory group then the nature of the advisory body is to give advice after deliberation, i.e. to provide advice. However, it seldom happens. PALM says "we take on board your comments". But our comments are for the Minister not for the PALM, and how come PALM filters the comments come up from LAPACs? What usually happens, someone [LAPAC coordinator] from PALM takes the minutes. This minute goes

to PALM and stopped at the planners' desks. How can the minutes be considered as advice?

Though the majority of the members believe that the Committee's role is advisory, the above comment indicates that their role is not advising the Minister but advising the planning agency. They are not able to advise the Minister directly, as all comments and advice usually stops at a level where it should not stop. A member stated:

We could play more proactive role in planning decisions provided our comments reached the right places, at the right times.

Some members feel frustrated about the outcomes of the Committee process and their role in the broader planning decisions. They are involved at a later stage of planning decisions and have no way to alter policy that already exists. The following comment of a member explains how he thought about the role of the Committee in planning decisions.

You have created an advisory body to give advice on planning issues; however, sometimes, you [PALM] come and say "this is what we have decided" i.e., the decision has already been made, then what is the fun of creating and taking advice from the advisory committees?

However, there is unanimous agreement among interviewed members that Committee is not, and should not be, a decision-making body. It should remain as an advisory body unless it is an elected body. They are just an advisory body, not a decision-making body with which some people confuse it. A long-time member noted:

Comments being taken into account or not taken into account in the planning process is a different case. But LAPACs have to look at the process to consult people and to get their input into the planning decisions. Consultation is part of the process. It is like a small window, with a bigger window behind.

However, some business representatives have been overtly critical of the role of Committee. One business representative felt that the role of an Advisory Committee in planning decisions was totally adversarial. He raised the question of members' understanding of planning procedures and noted that advisors in any planning Committee should be those people who had the capacity to advice. He also felt that Committee was not a body with such planning expertise but a Committee of people comprising residents, community activists, businessman and planning professionals. Therefore, some felt that the role of a Committee was always adversarial and conflicting, and believed that the prevailing motive was always to delay the consultation process, particularly DA approvals. Some resident members do not want to see any changes to their suburbs, while, some business representatives feel that there is a market demand for affordable housing in the suburbs near Civic. With the issues of development and redevelopment in the Civic area, there is always a confrontation between business representatives, residents and community groups. Residents and community groups interviewed posed the same views about their role in participation process. Since there was conflict between the residents and business groups on redevelopment work, there was hardly any consensus between them. They realized that their role was always adversarial in the meetings.

Technical officers' views on Committee's role

Technical Officers who usually attend meetings to represent the planning authority were also interviewed. All of them feel that Committee is a diverse voluntary group of people to make comments on planning and development matters, but there is a difference of opinion about the role of the Committee in planning decisions. Technical Officers are believed to have an adequate understanding of the terms of reference and the role of the Committee in planning decisions. When they were interviewed, they expressed very different opinions on Committee's role, and the overall consultation process carried out by them. Technical Officers have their own views on why Committees exist, how Committees provide advice on planning matters, what level of planning services they provide to planning staff as well as to the Minister for Planning, and what role they may possibly play in the future.

Some senior Technical Officers carry similar views. They recognize that Committee is a body providing free consulting services for the planning authority on various planning issues, but some often disagree with aspects of recommendations and advice. However, recognizing its importance in planning decisions and acknowledging its role as a group for expressing needs, expectations and preoccupations, some senior members found themselves dissatisfied with specific Committees and their members on planning issues. They noted that Committee members sometimes asked for more information, which planners believed to be an intentional move to delay the whole process. Two Committees, Manuka and Inner North, seemed to be more problematic as Committee members of these groups demanded unnecessary clarification and planning documents, which might not be required for the DAs under review. A senior planner saw this situation as frustrating and a waste of time:

> Some of the members know planning laws and know when to object to any application to delay it. It is done intentionally that developers would miss the market and move away from it [current DA].

Another senior member of the planning team also made similar comment:

Overall, it is good to have a forward body to consult on local issues with people living in these areas. Sometimes, their comments are not possible to accommodate in the planning process.

A business representative, who is also a professional planner, identified the causes behind not complying with the recommendations of the Committee in the planning process:

Sometimes, members come along with a planning philosophy that can fit only with 1930 and 1940 ideas of building codes and with veranda and terracotta, which is ridiculous to fit in the present day of high-quality architecture. It is because the members have no prior knowledge of planning and design aspects, which I think, require for making intelligence comments on design and planning process in the urban setting.

Other Technical Officers made similar comments on the role of other Committee members. They stated that Committee had been going beyond its major purposes as outlined in its protocol and had moved into the areas that normally are within the discretionary powers of the Minister for Planning. A PAC should be a body of people that conveys community concerns into the planning process, not a means of identifying planning matters, which is the main concern of planning agency, not a group or the community. Residents and community groups are on a Committee to advocate pertinent and overlooked issues that may have real implications for planning and development, but members are not in a position to dictate the planning process, which should be designed and implemented by the planning authority. Any problem encountered in defining the consultation process, can be solved using effective techniques. Another planning staff member supported this statement:

Their [Committee members] role is to view only community concerns that can come up with some naive comments, then we [planners] can say" yes, your comments have got some basis in real situations" and then we can tap those, as part of their advice.

He gave an example to justify his comments and clearly identified what role an Advisory Committee could play in the planning process.

Well, Adelaide Avenue [an avenue that links National Parliament to the south side of Canberra City] is degrading, and certainly some of the other members say—"yes, it is". Now, we can ask why and how it is degrading. Then we can get their views and identify the issues, which needs to be addressed by the planning authority. As planners, it is our concern to tap information from the community to put through the planning process. I therefore believe that their [LAPAC] role is to raise community concerns not to dictate the planning process. We have to remember that their views are coming from their heart and expressed emotionally, which is sometimes difficult to accommodate in the planning process.

Technical Officers were also concerned about cases where a Committee member writes directly to the Minister for Planning or Executive Director of the Authority stating facts in opposition to the continuing planning process. Even though everyone has the right to make submissions to the Minister stating concerns on development applications, the Technical Officers feel that this practice should not be encouraged because it would not show any supportive activities by the members on any development application. Most of the planners believe that Committee raises only negative concerns about development activities but never praises their economic benefits and greater development for the community.

It was observed in the interview that some planning staff did not see the Committee as citizen experts on planning issues. They, therefore, believed that Committee members would not have much to contribute to the planning process. In general, planning personnel feel that a Committee is a group of community activists who seek more unnecessary documents and irrelevant information from the proponents and the planning agency. A Technical Officer commented that members might not understand the planning documents or would not go through the necessary details provided by the planning authority.

Developers' views on Committee's role

Two developers, also members of two different Advisory Committees, were interviewed. They were represented as the businesspeople on the Committee, because there is no provision for developers to be a member. However, sometimes, they become Ministerial appointees to represent the business groups. Surprisingly, when approached to one of them for a formal interview, after identifying him as developer, his response was negative: 'I am against it'. When asked: 'against what?' We did not ask you anything, then, he replied, 'I am against LAPAC'. The author reminded him: 'You are a member of LAPAC, and then he said, I am against residential and community reps [representatives], as they want to stop development in their own suburbs'. The author replied that he was not part of them, but simply evaluating the consultation process and perceptions of all members about the process. Finally, the developer agreed to be interviewed.

He started very aggressively with a comment that total consultation process was useless:

> Sometimes they [residents and community representatives on the LAPAC] ask for a tree survey for the proposed development. It is not the job of

LAPAC, but they are entitled to ask the question. Developers say "no". As a developer [but representing business groups on the Committee], we don't have to give a tree survey to the LAPAC.

The above incident indicates that business representatives normally belong to developers' community and always try to promote the interests of development proponents. In addition, they are critical of the consultation processes even though they are members of the Committee. Business representatives seem to believe that Committee belongs to the residents and community groups, who see businesspeople and Ministerial appointees as just opponents of the residents and community groups, and feel alienated from the consultation process. The business representatives also feel that there is no feeling of equal partners in the planning process. This conflicting and antagonistic feeling also exists among the developers, who believe that asking for additional information is to delay the process of DA approval, and the delays cause the developers to lose business tenders and subsequently their income as well. A developer observed that Committee was overstepping its boundary by seeking legal power to be a part of the planning decisions. However, it has been noticed that there is a strong feeling that the Committees should have a legal basis to be a parts of the Administrative Appeal Tribunal (AAT). Developers do not agree with giving legal status to any PACs, which they believe would put another aspect of planning into the planning decisions. They also believe that the Advisory Committee is essential for communicating with the community but should remain in an advisory position to raise community concerns only, not to dictate the planning process, which should remain within the planning agency. A developer said:

> If this [legal power] is the case such as the LAPAC is given a legal base, you must have qualified people on the committee. It seems you are putting another arm of planning to the consultation process. On the other hand, if they [LAPAC] become part of a legal base, they can be sued by the developers. We had a case, where a neighbour lodged a complaint to the Court regarding topple a dual occupancy adjacent to his house. But he lost and incurred money to us.

Business representatives are found to be frustrated with their colleagues for seeking the legal power to be a part of AAT. A business representative commented that there was no such example of an advisory form of the consultation process where the Committee was given legal privileges. Only affected and concerned individuals, not a group like this should be entitled to lodge applications with the lawful authority. On the other hand, a community representative of Manuka Committee believes that if a community group such as *Old Narrabundah Community Action Group* can lodge an appeal with AAT, then there should not be any problem with LAPAC lodging an appeal to same lawful authority. He added:

If we have a consensus decision and authority topples our concerns, then we can go to the AAT like an affected individual and group.

However, when asked if they lost who would incur the cost, a Committee member commented that there should be a provision to deal this situation as well.

Wider public's views on Committee's role

The wider public were also interviewed across all six Committees. They feel that they have a good understanding of Committee's role in planning decisions and see a Committee as a source of information that is required by the planning agency to register community input into planning decisions. In general, the wider public feel that Committee is a good platform for public opinion and information and can make recommendations to the planning authority on behalf of the community. The interviews with the wider public indicated that members of resident associations and community groups were very pleased to see Committee becoming more active in getting public comments and residents' views for planning decisions that would be acceptable to the greater community.

Members of the wider public mostly attend meetings on issues that may have implications for their neighborhoods. They seldom make comments on development applications, which is a regular practice in meetings. Instead, they turn to the meetings on issues like group center redevelopment, multi-unit development and the Master Plan for the area, which may affect their neighborhoods. They often write letters to the Committee convenor stating their concerns on proposed development applications. Sometimes they just appear before the meeting to express their concerns. Two wider public of Inner North Committee areas who were interviewed, were found to be dissatisfied with the response of Committee to the appeal of the wider public on development issues, particularly proposals for multi-unit development. One such proposal for Boldrewood Street in Turner was lodged with PALM, and residents were very much opposed to multi-unit development in their neighborhood close to the Australian National University and CSIRO. A member of Turner Residents' Association (TRA) came to meeting with a proposal for alternative options that reflect most of the residents' views. During the meeting, he presented a plan that qualified as a residents' option for the development work. Since his first appearance, he had been twice to meetings but had never heard anything from planning authority about residents' options. The following quote illustrates his views on the consultation process to respond to communities:

> I am profoundly dissatisfied with the current process of responding [by PALM] to the community. I have been here twice and have not heard anything about our concerns. I don't know when the community will

hear that 'well, we considered your views and are not able to incorporate them into the planning process. Thank you very much for your efforts to come up with an alternative scenario'.

However, Advisory Committee, as community-based organization cannot do anything on behalf of wider community to respond with their own options. These Committees seem not playing important role for making planning authority accountable to the community. Similar examples can be found in various Committees. For instance, extension of RSL Headquarter to a park area of Campbell suburb, the development proposal for Ginninderra Lake Foreshore and the extension of Yarralumla Child Care Centre, are controversial development applications, regarding which residents and community groups come to the meetings with their own concerns and explain possible impact on their neighborhoods. In these proposals, the wider public were not satisfied with the role of planning Committee nor with the consultation process and the decision outcomes. They were critical of the Committee and commented that the group was not effective in promoting community concerns to the planning agency. It did not truly reflect community concerns, and often was not upholding community aspirations and needs. Some wider public commented that Advisory Committee failed to convey community needs effectively to the planning authority, and so they questioned the role of the Committee members in planning decisions. However, dissatisfaction among the wider public is not a regular phenomenon, as some are found to be satisfied with the process. A wider public in Majura area was pleased that he received an immediate response from a developer to an appeal about sun-blocking in the proposed dual occupancy development in his backyard. The initial proposal for the dual occupancy would have blocked the afternoon sun from his living room. When he objected to this during a Committee meeting, the developer lowered the height of the proposed design.

A member of the wider public was interviewed who used to attend Manuka Committee meetings. He was working as an architect with a construction firm and was interested in the consultation process. He thought that this Advisory Committee was not a good platform for the community to become involved in planning decisions, since it would provide only the illusion of input. To make the Committee an effective platform for public participation, he suggested that it should have a high profile in the community and should be encouraged by all planning policy community to solicit real input, but the Committee has no provision for this.

A member of the wider public in the Burley Griffin Committee felt that the Committee's role should be to influence the decision-making process according to the community's preferences, but Committees seem ineffective in influencing planning decisions. Many reasons have been identified behind such ineffectiveness. Some of these are structural imbalances, limited time to comment, not enough resources to get public input, and lack of adequate information for the members. In addition, there is too much bureaucracy and a 'rubber-stamp' approach to the DA approvals. If the Territory Plan permits developers to do something about land use changes, they do not care at all about community objections. Rather, they do it and remind the community about the Territory Plan and the DA guidelines. This is an instrumental approach to planning and for development, even though the community may have reasons to object. A member of the wider public remarked:

> If the DA follows development guidelines and complies with the Territory Plan, like it or not, it's gonna be approved. No matter, it is ugly or beautiful if it's followed DA's guidelines, it's gonna be approved.

The wider public believe that Committee has a very limited role in influencing planning decisions, but the Committee is very useful in finding facts and community concerns, which the planning authority may have overlooked. In such cases, Committee can play a fact-finding role in the decision-making process.

It was observed that most of the wider public of both Committees strongly felt Planning Authority was not active in protecting Canberra from *Kingstonization*[1], and as a result, members of the wider public were very frustrated with the process. They also observed the limited role of the Planning Authority in planning outcomes and felt that the proponents might play a stronger role in planning decisions, as well as the political parties which have their own planning policies and political interests.

The wider public in the Inner North Committee are involved in many resident and community groups in the North Canberra areas such as the Braddon Resident's Association, Turner Residents' Association, Reid Residents' Association, Save the Ridge Association, Community Council and the Watson Community Association. All members of these organizations expressed more-or-less similar views on the consultation process and described the existing decision-making process as exclusionary and inaccessible to the general public. They believe that the locus of the decision rests with the Technical Officers and members of the Design Review Panel (DRP), which is a kind of parallel body to give professional advice to the decision-makers. They also believe that planning policy is made by a number of Technical Officers and by the vested interest groups of insiders and outsiders belonging to the developer community, who meet behind closed doors to develop policy recommendations for submission to the planning authority. In this process, Committee is marginalized and used as a vehicle to give legitimacy to the consultation process that people are being heard and given high priority. A member of the wider public stated his idea about the whole planning process as:

> .... senior Technical Officers meet privately with developers and businesspeople to hear their concerns and ensure that new policies are in their favour. They [Technical Officers] are not planners; they are

pro-development staff. Planners are those who are able to balance between community and proponents. But they are on the other side.

This respondent pointed out that planners were never on the side of residents and community groups when discussing a development application. It looked as if development applications belonged to both the developers and the Technical Officers. This member of the wider public believes that more housing units, more dual and triple occupancy and occupying the parks become the main planning objectives of PALM. The planning directions promote the loss of uniqueness and the Bush Capital image if development continues, so Canberra needs to stop development work without proper planning guidelines, in order to protect its garden city image from the 'scatter-gun' planning approach. He further added that Canberra needed a development control plan (DCP), which was also reflected in Taylor's comments. Taylor (2001a) defined the 'scatter-gun' approach as development, occurring haphazardly where the opportunity occurs (Taylor, 2001b). Taylor also identified various drawbacks of the planning process in Canberra and termed the approach as one, where developers felt the need to redevelop, while the planning authority responded only to their proposals (Taylor, 2001a).

Though the West Belconnen Committee covers twelve suburbs, it has not faced much development or redevelopment pressure. These suburbs are in far north of Canberra, along the NSW border, and depend mostly on Group and Local Centers. The planning challenge for this Committee was the about redevelopment of 'declining' Group and Local Centers, and to formulate policies which would provide opportunities to develop each Center to its potential while maintaining appropriate standards of development and environmental quality and a balance of sustainable Centers. There was a need to change land uses of Local Centers and to find alternative ways for their viability, which required variations to the Territory Plan to allow changes in existing land uses. PALM proposed a mixed-use redevelopment plan for all Local Centers, but the residents and community groups of these areas were not satisfied with the decision to go ahead with mixed-use development of the Local Centers. The wider public were profoundly frustrated about the consultation process. One member of the wider public in the West Belconnen area commented as below:

> Developers in the first instance go to see PALM. He [developer] comes up with a proposal and PALM gives some guidelines and says, "Well this is what you have to do and comply with until or before you go to see the people". So, they have developed some strategies where they have to say "You have to inform the public and let them know". What is happening is, instead of leaving it to developers, PALM and the developers come together and give the impression that PALM is in the pocket of the developers. And PALM says "That is what we are doing for the community, this is good for you, should have it" as if they knew what we wanted.

Overall, the public expect Committee to act as a watchdog to ensure that development applications are not moved quietly through the bureaucratic and political process without being rigorously reviewed by interested residents and the community. Since the wider public have no vote in the Committee meetings, they feel that Committees must scrutinize the activities of the residents and community associations to take advantage of the input mechanisms available. Although the wider public often feel they are being ignored and not being heard by planners, they acknowledge Advisory Committee as a way for the community to provide necessary input into planning decisions. They also state that Committee's influence on planning decisions is hampered by the lack of fairness and openness in the consultation process.

Some of the wider public have very positive hopes for Committees roles and their outcomes. When the Advisory Committee was established in September 1995, they had hoped that the Committee would take an essential part in upholding the community's aspirations and expectations. Some felt that their views on planning and development matters would probably be channeled through Committee's activities and would be considered in the decision-making process. When the Committee became operational and advised the planning authority on development applications, the community started to feel that the Committee had not systematically and effectively conveyed their expectations to Planning Authority. They identified various drawbacks of Advisory Committee and would like to see its role increased to cover a wide range of planning issues. The wider public also felt that the Committee protocol prevents the advising Committee from contributing effectively to planning decisions by excluding various planning issues.

The protocol states that the main role of PAC is to advise the Minister for Planning on planning and development matters. However, interview indicated that there was no consensus among all members on their roles and the working mandate of the Committee. The protocol has been evolving over the time into policy-making, because the members have added to their meeting agendas issues of concern such as dual occupancy, which Committee protocol had excluded with an explanation that the Committee was overwhelmed with many development applications, so it might be difficult to complete its main task of advising the Minister. Therefore, dual and triple occupancy matters were not included on meeting agendas, nor were other planning issues such as transportation planning, ACT Access Planning for disabilities, areas under commonwealth control, land release, land reclamation and land sales.

Some members feel that if Committee wishes to become more active in the overall planning process and decisions, it risks losing its focus and may not be able to address the main planning issues. It may also find it difficult to manage with the limited time and resources available, which the government in the ACT advocated excluding dual occupancy matters from meeting agendas. Some members want to expand their role, not only advising on DAs but also collecting community input. They want to include all planning matters on their meeting agendas and to see the current protocol amended to include all planning-related issues in the terms of reference.

Other members, particularly business representatives, believe that expanding their role would take more time than they have been allocated, and more time is not available. If the Committees were given more to do then it would be difficult to complete meetings in time. Some members express enthusiasm about increasing the number of meetings to at least twice a month, considering the large number of development applications, but business representatives disagree with increasing the frequency of meetings and feel tired of the consultation process. A business member commented:

> As a businessman, I cannot expend more than two hours in a month. However, currently, I spend more than that to oversee the current DAs for the meeting.

6.2.2 Evaluating effectiveness: Promote learning

Kweit and Kweit (1981) pointed out that public participation in the government decision-making process improves education and knowledge for citizens and bureaucrats involved directly or indirectly in participation. Webler et al. (1995: 443) commented that public participation, which includes deliberation and inclusion, 'can initiate social learning processes which translate uncoordinated individual actions into collective actions that support and reflect collective needs and understanding'. Arnstein (1969) observed that an effective participation process would enhance the participants' understanding of the planning process and decision outcomes. In addition, regular outreach activities and orientation programs for the participants on planning issues would also help participants understand planning and development matters (Kweit and Kweit 1981). Understanding planning matters allows participants to carry out the role envisaged in major planning laws involving identifying violation or compliance by the development proponents or development applicants by applying community pressure and by enforcing laws on any development and redevelopment proposal. For instance, a greater understanding by the Advisory Committee members of the Territory Plan and relevant acts and regulations is essential for discussions with the development proponents on many planning and design aspects, such as maximum height limits of a building in commercial or residential areas, plot ratio, allowable dual occupancy on big block, available open space, allowable land for other commercial developments, unit titling and front and rear setback. Webler (1995) commented that a good understanding of the issues on agendas and approval process of the project is an effective criterion to evaluate a public participation process. Understanding various planning and design laws would give ordinary Committee members a great deal of planning knowledge to deliberate issues and formulate alternative options. Lauber and Knuth (2000) pointed out that if citizens participate in the decision-making process, and if one of the goals of decision-making is to produce high-quality decisions, some efforts to educate or inform citizens should be inevitable. Therefore, educating the participants has become an important criterion to evaluate an effective participation process.

This book discusses two aspects of this criterion. The first aspect discusses participants' perceived understanding of planning matters involved in the consultation process of Advisory Committees, while the second aspect discusses whether the participants actively feel that they had enough knowledge of planning to contribute effectively to planning decisions.

In the case study, PAC has not been successful in increasing participants' understanding and awareness of planning issues. After serving over a period of years or so, attending scheduled meetings, and going through huge amounts of planning documents, Committee members commented that lack of educational and outreach programs made them unable to contribute effectively to planning decisions to preserve the uniqueness of Canberra and its image of Bush Capital. They identified various reasons for their ineffective contribution to the planning decisions, mainly their unfamiliarity with the processes that are normally carried out through Advisory Committees and with the Committees' main function to advise on planning and development matters. As a result of their unfamiliarity with the Authority's planning process, Committees sometimes spend a great deal of time in learning about planning rules and design guidelines in order to become familiar with relevant planning process, most particularly with the process of development approval.

New members were also not given prior understanding of the planning process and the essential steps of development approval, which was necessary for them to make comments on development applications. Despite this, they tried to make themselves familiar with the planning process, but unorganized planning documents and excessive use of planning and design terms prevented them from understanding the planning process in timely manner. Interviews with other members indicated that they had no adequate understanding of development applications, except for a few who had long been involved with the process.

Most importantly, it was observed that only a few members had a working knowledge of the Territory Plan and the *Land (Planning and Environment) Act 1991*, (now *Planning and Development Act 2007*) and most of the time, other experienced members had to explain their necessary clauses and features. The variations to the Territory Plan are not also clearly understood by most of the Committee members. Without prior understanding of planning rules and acts, it would be impossible to contribute effectively to the proposed development programs. Some resident members who were interviewed feel that a PAC does not need to be comprised with experts in the planning field, but they acknowledge that some working knowledge of the planning process and relevant laws would definitely promote the understanding of planning and its consultation process. However, a former member, by profession an architect, evaluated the consultation process as 'totally waste of time' for him, and commented that all members should have professional and construction knowledge in order to make informed comments on planning matters. He had found in the Committee that this was totally absent in most of the members, so he decided to step down from the Committee.

Some resident members were critical of members who were motivated by political ideologies and tried to dominate the discussion. Some local political leaders, who became Committee members, used it as a platform for their own political parties. A convenor indicated that politically motivated members had no knowledge of planning and construction matters, but they tried to make comments as if their option was the only correct and accurate opinion to be accepted by all. A former member who described the reasons for his resignation made similar comments:

> I didn't stand for re-election, because there was no expertise in the panels and the people who got there. I think most important issues were overlooked. People have their own pet hobbyhorses about things.

It is, therefore, evident that the PAC should have competence in processes that enable everyone on the Committee to become educated and aware of planning and design matters. Another former member felt that Committee was deviating from its own mandate of dealing with planning and development aspects and that the Committee's sole task had become commenting on design aspects, not on planning matters:

> Though planning and design aspects are indispensable phenomena, LAPAC is mostly dealing with design aspects and people got there making comments on design issues and think that these are planning problems. Some of the people are trying to become designers instead of planners. We are actually looking at the design problems rather than the planning problems.

This respondent added that many people were acting outside their field of expertise. For this reason, he found that it was a waste of time for him to continue and so resigned. He believed that the Advisory Committee should be formed by a group of people having knowledge of planning and design matters:

> If the reason of the LAPAC is to get community input and you want the layman's view, then it is fair enough, but if you want informed comments, then you must have people with knowledge.

Another member, who resigned, was critical of the Technical Officers' level of competence understanding planning process, while he was simultaneously critical of the other Committee members as well. The following reflects his concerns about the competence of Technical Officers:

Some committee members have knowledge and experience that is greater than PALM representatives. When the members may ask questions, the PALM representative is not competent enough to answer or does not have relevant experience to answer those questions. All they [PALM] have is some inexperienced young professionals.

The above discussion has given an indication that the consultation process does not adequately facilitate educating the Committee members as well as the wider public, but respondents saw the importance of introducing various planning laws and *Acts* to Committee members. Without a competent understanding of planning laws, guidelines of high-quality sustainable development, and the process of development approval, it is difficult for them to contribute effectively to the planning decisions, so as to give competent advice to the Minister for planning. A long-time member said that Committee should promote learning objectives, such as educating its own Committee members and educating the wider public outside the Committee. The existing experience and knowledge of members combined with continuing face-to-face discussions and access to planning documents is likely to be educational for the members. To educate the public, the planning agency should organize outreach activities, such as planning workshops for the wider public.

One of the authors asked all the Committee members who were interviewed about the level of understanding they had reached on planning matters through their participation. Most of them had gained understanding of a wide range of community views and the political processes in development approval. Some architects and planners on the Committee said that they already understood the planning process and development approvals, but they acknowledged that a wider range of community views and relevant concerns on planning had much value for planning decisions. They believed that community concerns should be considered for greater community satisfaction. There were reports of increased individual understanding of development applications with less confusion about the process of development approval. The long-time members from all Committees reported a greater understanding of the planning process and the process of development approval. A few perceived themselves as having more confidence in the planning process, who were mostly people with professional expertise in planning. Some professional planners reported that they had enough expertise in planning and design matters so they did not require any additional program to improve their planning understanding. However, they felt that residents and community groups had inadequate understanding of planning, and therefore should not be included on the Committee to give planning advice. They agreed on the importance of having representatives of the community on the Advisory Committee but acknowledged that selection criteria should be formulated in such a way as to include people with planning knowledge on the Committees. Some architect-planners on the Committee commented that a few members of the Committee were unable to understand development applications, and could not even properly understand a map and its features, but continued to make comments on plans normally prepared by a group of professionals in fields such as architecture and urban planning. A member commented:

> These people [LAPAC members other than business groups and ministerial appointees] don't understand how to read a plan, they cannot understand the difference between section and elevation, don't understand about Master Plans, don't know about contour points, and even cannot find the North point of the plan.

A business representative made similar comments. He believed that it was not the other members' responsibility to understand planning laws and the high-quality design guidelines. He thought that information provided to the Committee members was well documented, but some members were unable to understand it clearly due to their inability to read and understand a plan. He commented that planners were not responsible for writing an ABC of planning for the other members on the Committee to make them understand. It is a long and a tedious learning process.

6.2.3 Evaluating effectiveness: Communication

In order to carry out its advisory role effectively, Committee requires adequate access to the information, and timely communication with the planning staff and other development proponents. Committee also needs to have contact with the residents, community groups and wider public to get their views on planning and development matters. However, interview results demonstrate that Committee does not have formal guidelines for communicating with planning staff, development proponents and wider community organizations. Communication in the whole Committee's consultation process is unstructured and not supportive in sharing and exchanging planning information. Some regular members who were interviewed feel that senior Technical Officers are not always cooperative and approachable when related information is needed on development applications. This lack of open communication limits the opportunity of some stakeholders to contribute effectively on development applications preventing them from participating in overall planning decisions. As a result of poor or suspicious relationships between Committee members and Technical Officers, there is less effective communication and less timely feedback that the community wants.

In the absence of any formal feedback system, Committee members are usually unaware of the comments and recommendations that are considered in the planning decisions. There is an informal system to let the Committee members know about the decisions made by Planning Authority, but correspondence between them is often very slow. And it was reported that all members were provided with a bare minimum of information. As a result, the Committee members receive no details about the discussions and the acceptance or rejection of their recommendations. The Committee members interviewed observed that communication with Planning Authority was better when the Committee included more business representatives, who were believed to have a visibly close relation with the planning staff. A community representative feels that communication between business representatives and planning staff assists the Committee in determining recommendations in a way that Planning Authority finds acceptable, as they believe business representatives normally look after developers' interests in the meetings.

The interviews with the wider public and Committee members indicated that the communication between them was quite frequent and regular. This communication includes regular attendance at the resident and community groups' meetings before scheduled Committee meetings, informal conversations, phone calls and emails. While the communication and feedback links between the Committee and other residents and community associations is very effective, some members feel that communication with other interested persons and community groups could be improved by providing them with meeting agendas and minutes. Committee convenors could play effective role communicating with others, but they do not communicate effectively by sharing common planning information and making relevant comments on development applications. A resident member commented that the convenor of her Committee was not regular in responding to the emails sent by residents or community activists. She believed that the convenor did not read emails and did not feel it necessary to communicate with Committee members, letting them know about important planning events involving their neighborhoods. She added:

> There is a workshop on 'community needs assessment' [organized by PALM], as LAPAC members we have not been briefed before, not being informed by the convenor. Why should I type all the news from the *Chronicle* [a local newspaper] and send it to other members? It should either be PALM or the convenor to let us know about on-going planning activities. But we hardly heard about it.

It appears that there is general dissatisfaction with the methods of communication between Committee members and planning staff. Resident and community groups often send emails to Technical Officers asking for general information on current development applications, but it takes a long time to receive acknowledgement. Even if Technical Officers do reply to emails with a further tentative date to provide the information, this too comes with very little explanation. However, Technical Officers do not agree with the statement of the Committee member. A Technical Officer said:

> Sometimes, I receive 20 emails from a resident representative asking to provide info, how can I do that? They have to come to the PALM shopfront to find documents ... *Land Act*, guidelines of HQSD. I cannot read

the *Act* and give explanations over the phone. It is their responsibility to know the relevant *Act*.

Another Committee member interviewed acknowledged that there was a gap between Committee members and the planning authority in sharing planning information:

> Mr. Clinton [not the actual name—a Technical Officer] is not cooperative or friendly to the new members communicating with each other. He hardly pays any attention to the members who want documents or survey reports with verbal explanations.

It has been noticed that members of the Manuka and Ginninderra Committees are very active in communication with other planning policy community. The convenor of Manuka Committee maintains regular contact with Committee members to share information on important planning and development issues. Members of this Committee have internally a good communication, and with the convenor as well with the Committee coordinator. The convenor also has close contact with the coordinator and praises the coordinator who works consistently for the Committee and is very regular in responding to the queries regardless of their relevance to the development applications or related matters. However, the convenor is disappointed in communicating with Technical Officers on sharing planning information. A convenor interviewed commented:

> The coordinator is very regular, because she does not have any value attachment to any DA, however, TOs [Technical Officers] are not regular with our questions. He [a Technical Officer] often takes a very long time to respond and changing TOs one after another also creates problems for us effectively communicating and understanding.

There is no specific or assigned Technical Officer for Committee meetings. Rather, it appears that whoever is free on that day can attend meetings to represent Planning Authority. A long-time member commented that this was another important issue, resulting in ineffective communication between Committee members and planning staff. All the members interviewed were unhappy about the variety of Technical Officers to Committee meetings. They feel that changing Technical Officers regularly makes a gap between Technical Officers and Committee members in communicating each other.

Residents and community representatives on the Committee have regular contact with each other. Some of them have various community relations and involvement with residents' associations, community councils and other planning and environment related organizations. All of them normally receive emails from a listserv discussion group maintained by a Committee convenor. The Manuka Committee has a single email discussion group from which all messages are directed. Apart from the individual correspondence of the Committee members, the convenor sends emails that contain community news to the local publications *The Belconnen Chronicle, The South Side Chronicle, Jamo, Narrabundah Pride* and *TRA E-news*. This convenor also sends community news to all members with Internet access either at home or at work. Most recipients of *E-news* are neighborhood residents and Committee members. The E-news also reaches out to planning staff, the ACT government officials concerned with relevant development applications and a few NGOs that work on planning and environmental issues in the ACT.

The newsletter is sent periodically or monthly, and includes information on forthcoming association activities or meetings, the meeting agenda and development actions and planning decisions that affect neighborhoods. The Committee, as a formal statutory body in the consultation process, has no newsletter to make the community at large aware of their agenda, but there is a regular notice and advertisement in *The Belconnen Chronicle* about important topics to be discussed at the forthcoming meeting. A member commented:

> A large number of the community is being left out from the LAPAC agenda, because LAPAC activities have never been sufficiently advertised. Only concerned members, politicians and activists look at the *Chronicle*.

Effective communication: Visualization of spatial information

One of the important issues in planning consultation is that participants have to make comments on the draft plans, which normally come with various spatial information, maps and figures. Participants in a consultation process often have difficulty in understanding spatial data, maps, or other planning documents, which may be in the form of digital or paper products displayed on maps and plans. Miscommunication and complications in understanding of spatial data can lead to mistrust amongst planning stakeholders. An effective participation process should therefore be one that uses spatial information in a way that participants can understand easily so they can make informed comments on draft plans. This book examined how the planning authority and developers used spatial information and provided it to the planning stakeholders, and how it was perceived by all stakeholders. This was examined through attending meetings and observing how development proponents used geographic information and technologies to make planning stakeholders aware of their neighborhoods, communities and the spatial features of the regions. All members interviewed were asked about how the planning authority provided spatial information and how they perceived this.

Communication with the planning policy community is often established discussing spatial information on planning documents, so visualization of spatial information given to the concerned participants has become an extremely important part of the planning process. Driel (2001, cited in Ghose and Huxhold, 2002) claimed that 50 per cent of the human brain's neurones are involved in vision, and 3-D displays can stimulate more of these neurones and hence involve a larger portion of the brain in the problem-solving process. 3-D computer models can thus stimulate spatial reality, which will allow viewers to recognize and understand changes in elevation more quickly. In addition, visualization tools enable viewers to see how new land-use and other policies can change the environment and physical structure of neighborhoods (Ghose and Huxhold 2002). More importantly, the visualization technique can show the outcome of a proposed development program and assess its desirability before the development work begins. Regarding visualizing spatial information in the meetings, Committee members commented that this was the most inefficient and ineffective part of the consultation process. They often had difficulty understanding spatial information displayed by developers and in many cases by the Technical Officers. Their frustration often leads to miscommunication and mistrust among the planners. A member asserted:

> It is rather difficult for us to understand what the building height of a DA is and front or side setbacks allowed by the Territory Plan. Maps and drawings are far away from us. Who will go there [where it is stuck on the wall or hung up] to see their [developers'] maps and drawings? Only Clinton [not real name] goes too close to the map for understanding. I'm nearly 75 and cannot walk up to there.

It is evident that with an efficient visualization process, the participants in a consultation could experience and visually understand the effect of proposed development programs in an intuitive and interactive way, and so be well informed about the possible effects on the neighborhoods and their surrounding environment. This could lead them to produce a community vision that would possibly be adaptable by the planners for final decision-making. Talen (2000) commented that urban visualization with any forms of multimedia could be a valuable tool for designers and urban planners. The ability to visualize potential modifications to the urban fabric and their actual context would allow planners and designers to evaluate alternative options rapidly, in more detail, and for lower cost than through more traditional analysis. The introduction of visualization would demonstrate the results of planning process visible to those concerned with planning decisions and allow the wider public to view the proposed changes to their environment in a realistic manner (Talen, 2000).

The importance of using geographic information and 3D models in public participation process has been acknowledged in the planning literature (Ghose and Huxhold 2001). All participants in planning consultation processes make direct or indirect comments on the spatial information that comes in the form of development applications, so communication and understanding of maps and plans are of great importance for making useful comments. Plans consist of geographical information and spatial descriptions about a place and its land use, so spatial understanding is an essential part of planning consultation models, without which it is rather difficult to contribute effectively on any development application.

In the case of Committee meetings, every DA and Master Plan come with maps and architectural drawings, and members have to comment on their merits in considering existing land use practices. Accordingly, developers come to the meetings with the proposed DA and its various features in the form of maps and drawings. A copy of all planning documents is normally provided to the committee members representing a particular neighborhood. Some resident members who were interviewed were dissatisfied with the maps and drawings provided, and with the way the spatial information was displayed and presented in the meetings. A resident member commented:

> The architect sticks the map far ahead of me, I hardly see anything there. Without details of maps, it is rather difficult to communicate about the contexts of development applications.

Members are greatly concerned about the use and display of geographic information. They prefer to see regular use of 3D models in presenting development applications, whether big or small. Some developers have presented development applications using 3D models, which Committee members believed to be an effective way of communication among the planning policy community. Some developers who were also business representatives were strongly critical of other members' competence in understanding spatial information. A business representative commented that some members could not point to the north point of a map, so he felt it was not worthwhile to introduce geographic information systems at the meetings. He was totally opposed to the concept of PACs that included diverse representatives of the community.

Some regular members are critical of the way the geographic information is displayed and presented in planning documents. They noticed that poor use of geographic information on plans led to ineffective communication between planning staff and Committee members. A member commented:

> The development proponent sticks a map on the wall with the help of a Technical Officer. The hand drawings sometimes leave out the patterns and location of various attributes. We cannot concentrate on one map as they [developers] keep changing one after another. They should have explained clearly each and every one.

However, there has been high praise for particular developers who always provide 3D models for presenting development applications, but this does not often happen, so it is difficult for members to become familiar with the maps. Developers disagree with the comments made by Committee members:

> We normally present a DA more than one time before the LAPAC members. Initially, they might have difficulty to understand, however, the next time they should understand.

It appears that developers are not very satisfied about explaining maps and plans to the Committee members. They blame each other. One convenor commented:

> I've to go very close to the map to understand. Others are reluctant to come forward to see the map clearly.

This statement indicates that Committee members are concerned about visualizing spatial information in a meaningful and effective way to communicate with each other. The blurred and inadequate information displayed on maps prevents the Committee members from making comments on DAs and Master Plan. A member raised the necessity of providing every member of the Committee with the neighborhood indicators about the areas of concern including pertinent information on demographic features, community facilities, housing, land use planning, commercial and retail space and transport and traffic conditions. Interviews with resident members indicated that most of the members were not satisfied with the information provided through maps and plans.

All members were asked to evaluate the use and visualization of geographic information in support or clarification of the development applications. Most of the resident and community representatives interviewed agreed that it was difficult to understand all planning terms used in the maps and drawings and felt that insufficient information was provided in displayed maps and plans. They were also concerned about continuously changing maps in consecutive meetings. Normally, the Committee ask for modifications in proposed development applications, if they feel modification is relevant to the betterment of the community. Accordingly, developers produce changes to aspects, which Committee members have not even proposed to modify. A member said:

> Sometimes we are lost when we see the second stage of maps, where little changes are made, which we cannot compare with the initial one.

Moreover, developers hardly give any additional information in the form of maps of areas surrounding the proposed development applications, so it is difficult to understand the whole situation at a glance. A Committee member commented:

> Say B13^2 allows a nine-storey building, but we don't see information on nearby areas whether it is B11 or B12. If we see them together, we could have given comments on over-shadowing.

The business representatives on the Committee have different views. A business representative of Inner North Committee commented that every member of the Committee should have a clear understanding of ACT land zoning and a working knowledge of the Territory Plan, which is the land use plan for the ACT. He added that it is not the developer's responsibility to make them understand about land zoning in the area surrounding a proposed development application.

It appears that DA presentations normally show the existing land use of the areas. The development may have a greater impact on the adjacent areas too, but Committee members often do not receive maps of surrounding areas to get a broader view of a development application. Developers do not give the details on adjacent areas. A resident member of Ginninderra Committee commented:

> I have reason to believe that developers do not come intentionally with sufficient maps and plans to support their DAs, in order to avoid a scenario which might change their original plan.

Another member of the Ginninderra Committee commented that they were given geographic information without details of a neighborhood. One Committee convenor indicated that this information was required by the Committee to understand how to improve the neighborhood housing conditions in the areas where redevelopment pressure was very high. He added that:

> We are not given maps that show housing conditions declining and housing stability and not given information at any time about numbers boarded-up and vacant and the numbers of absentee rental properties.

It appears that Committee members are not able to effectively monitor the condition of their own neighborhoods and others. The absence of relevant geographic information for the participants impedes the assessment of changes in neighborhood conditions and to evaluation of the quality of development and redevelopment activities. A member of Majura Committee commented:

> What we need is a comprehensive property database with user-friendly software. In addition, we also need a Territory Plan that clearly indicates relevant rules for my areas.

6.2.4 Evaluating effectiveness: Relationships and trust

It was observed that the relationship between the Committees and Technical Officers varied notably, depending upon the Technical Officers in question and the planning and development matters at hand. A long-time member of West Belconnen Committee described the relationship between Committees and Planning Authority as both positive and negative. He stated that 'the

Technical Officers are sometimes cooperative, very helpful in responding to questions, and fairly professional in explaining planning and development issues; but at times they are adversarial and antagonistic'. It was also observed that the relationship between Committee members and planning staff was not always confrontational, but sometimes discussions on the development applications ended up in heated debate. Some members feel that there is still a very amicable working relationship among the planning policy community, and Committee is a way for the community to express their neighborhood concerns. Other regular members comment that the planners are trying to control the overall consultation process and also control the planning agendas to be sent to Committees for their contribution. A former convenor agreed that the planning authority sends development applications for community input at the end of the process in order to get approval within a time frame. The relationship deteriorates when members feel planning authority intentionally delays in sending development applications in order to avoid greater opposition from the community. One community representative stated that he does not want to 'go head-to-head' against the Technical Officer, but felt frustration when he realized that an application was designed to get approved without much obstruction.

However, all the members interviewed feel that the Committee coordinator and support staff at the front desk, including the director of the Territory Planning Branch, are very helpful and supportive to the Advisory Committee and its activities. But members have some reservations about senior planners, who they feel are not always supportive or approachable in providing necessary documents.

The interviews with former members indicate that some Committees do not have a cooperative relationship with others in the planning policy community. The poor relationship among the development proponents, the resident and community representatives, the business representatives and the Technical Officers has been deeply rooted in the context of the consultation process. As discussed, there is conflict among these groups in the way they see their roles in the public participation process. Committee believes that its role is to be active in bringing forward planning and design issues to the planning authority, while the business representatives feel that Committee is going beyond its terms of reference. On the other hand, the Technical Officers feel that member of the Advisory Committees are not qualified to dictate the planning process but is only eligible to raise community concerns. As a result, there is a very little effective collaboration among these groups, which limits the opportunity for the Committee to contribute effectively to the planning decisions. Burley Griffin, Manuka and Inner North Committees provide a good example of the poor relationship within these three Committees. According to some members, the antagonism has increased; former members of many Committees declined to stand in the next election to renew their membership partly because of the continuing animosity among them.

Some developers have similar views about the Committee. One developer commented that he was overwhelmed by emails from activists and felt that responding to their queries was a waste of time. He further added:

> If activists want to know something that I know, I should say, "go to the uni and do some planning courses as I did".

Overall, developers' views of the consultation process are not very positive. They see it as wasting time and delaying approvals for necessary developments. One developer commented that Committee should be commenting at the strategic level of planning matters, not on every issue-specific design and policy matters, which should be the concerns of planning professionals and design experts.

Resident representatives are far more concerned about redevelopment than are other representatives on the Committee. There are two polarized views with 'no middle ground' on different development and redevelopment issues. Sometimes there is no agreement on issues to which the Committee members give different values. Resident and community groups mostly oppose the multi-unit development in the inner-city area whereas business and Ministerial appointees support any development applications that propose to build multi-units in the suburbs. This conflict often occurs in the meetings and Planning Authority must explain the Territory Plan, existing land uses and possible variation to the Territory Plan. It appears that the meeting is a place for giving personal views on multi-unit development, and what is liked or disliked. Always there are two views: for and against. One Burley Griffin Committee member who was interviewed commented:

> This is entirely a process of making personal comments only, not making any collective decision. Some are supporting; some are not.

Another member of this Committee described this situation as purposely designed by the planning authority not to succeed. In his opinion, Advisory Committee as a group, does not make any decision, nor gives advice to the Minister, nor suggests anything. It simply talks about development applications and the coordinator just records the comments. This process has given Planning Authority relative autonomy in making decisions without any collective advice as suggestions from the community.

The Convenor of Ginninderra Committee noted that the formality of meetings needed to be increased to ensure that the meetings ran as smoothly as possible. At one meeting, some issues arose concerning the role of Technical Officers and presenters. The Convenor pointed out that problems would arise for the Committee if developers did not want to attend with DAs in a scheduled meeting; equally, DA presenters should understand genuine and complete consultation. Ideally, presenters should attend to answer questions and provide information to all members, not try to convince people of a particular view:

> I have come across some developers who want to point out only those questions to them, which they find comfortable and convenient to answer.

This convenor observed that some presenters were better suited to the consultation and that Committee would need to deal, in a formal way, with those who were not. He suggested that members should remain calm in discussions and that he might intervene to steer the meeting if he felt it was moving off the track:

> I have experience that sometimes we talk with very unnecessary issues and express personal feelings towards other members. This might not relate to the current agenda.

The analysis of this case study showed that all Committees were having difficulties in their relationships with the factions. These difficulties limit the ability of all members to fulfil their role effectively, as they generate an unpleasant and hostile atmosphere on many occasions. A poor relationship with Technical Officers has at times resulted in the diversion of essential energies from the task at hand and contributed to bureaucratic backlogs and delays.

The Inner North and Manuka Committees also demonstrated that the Committee operates in an adversarial and confrontational environment and there is a general unwillingness to cooperate among the proponents of differing views. The Technical Officers interviewed often feel that the Committee is overstepping its mandate and is naïve in the planning process, which creates trouble and additional work for both Planning Authority and Technical Officers. Some Technical Officers do not feel that Committees should have any role in planning decisions, or that planners should be accountable to the Committee for their decisions. Therefore, despite having a polite atmosphere, Technical Officers often feel that the PAC is overstepping its boundary and infringing on the responsibilities of the planners.

6.2.5 Evaluating effectiveness: Objective driven

One of the important criteria to evaluate the effectiveness of a participation process is to analyze whether the process achieved the authority's objectives, and the level of participants' satisfaction. Octeau (1999) notes that authority, if not independent, normally implements the objectives of a political party and its planning policies. Other interest groups consisting of community-based organizations, enthusiastic individuals, planning professionals and political parties, may also have some defined objectives to achieve through a consultation process. The benefit of evaluating the authority's objectives is to understand whether the objectives are met, and to identify the areas of concern that need to be addressed, which is believed to lead to formulating the necessary remedial actions for a better process. The authority may find that the existing processes do not require any adjustment, but other interest groups, either collective or institutionalized interest groups (Lightbody, 1995), may find it differently. Thus, the views of interest groups and participants become increasingly prevalent in determining the effectiveness of a participation process.

The main objective of the establishment of PAC by the government was to engage the community in a more coherent and transparent way, so that the community was heard and given importance in the planning decisions (PALM, 2000a). Similarly, participants have pre-conceived expectations from the consultation process of being heard and having adequate opportunities to put their concerns into the planning process. The book discusses the satisfaction level of all Committee members interviewed along with others including political leaders, town planners, community leaders and academics and examines whether their expectations were met through the consultation process of PAC.

Participants' satisfaction

All Committee members who were interviewed were asked about their overall level of satisfaction with the process of public consultation. In general, members expressed a moderately low level of satisfaction. However, dissatisfaction varies across the Committees: in the Manuka Committee, all resident and community representatives are dissatisfied, but Ministerial appointees and business representatives are not. Ministerial appointees and business representatives were found not happy with many of the representatives from resident and community groups. Inner North and Burley Griffin LAPACs also express high level of dissatisfaction on the overall consultation process.

The main city and its adjacent suburbs are in the area of the Inner North neighborhoods. The majority of Inner North Committee members are dissatisfied with the high-density development around the City Center. Residents and community representatives in Inner North Committee commented that the planning authority is deviating from preserving the National Capital as a 'Garden City' and 'Bush Capital.' They also believe that when self-Government started in this Territory with the Liberal Party, the party started selling land to developers. However, some business representatives on the Committee have different opinions about high-density development in the City area and believe that high-density development and urban consolidation are inevitable demands. A business representative commented:

> There is market demand. Developers are responding to the demand to provide services to meet present-day demands. There is pressure in the

Civic area for multi-unit developments. For these reasons, we are developing multi-units. It is our duty to quantify the demand of the present market. We consulted the market and found the demand for developing high-rise building in the central business district.

Resident and community representatives interviewed acknowledge that there is still a high demand for dwellings in the areas close to Civic, but there is enough land available in the Gungahlin area, so developers can develop there. Residential and community representatives are not always in favor of high-density development in the areas close to the Civic. A convenor stated:

We are not doing any good things. Only stopping bad things, not promoting good things.

It was observed that those who are mostly satisfied with the present process of consultation primarily represent business groups on the Committees. The interviews show that the most satisfied persons are either business representatives or those appointed by the Minister. Resident representatives from all the Committees show a high level of dissatisfaction.

The planning spokesperson of the Australian Labour Party also made comments on the consultation process and evaluated that PAC was established as token gesture to consultation. He commented:

The Variations of the Territory Plan by and large are not driven either by LAPAC or PALM to improve or change the land use in the particular location. Instead, it is driven by the individual requests of development proponents who decide that they want to do some sorts of development in the particular location. LAPAC has no role but only making comments on decided development applications. So, it is driven by the individual developer rather than more strategic and holistic views of what the demands are to change the land use in the area and how they meet community needs.

The DA approval process through the Committee appears to be nothing more than a formality, in which business representatives normally support the DA (in some cases DA lodged by their own office) and come to the meeting to have their private interests served. They are therefore satisfied; but perhaps, at the expense of other legitimate community interests which carry less weight in the planning process. Committee members revealed that their goals were not met through the Committee's consultation process.

Those who expressed dissatisfaction were examined more closely during face-to-face interviews. They were found across all Committees. Most of the former members who had resigned from the Committee expressed their dissatisfaction with the consultation process and with its implementing body, the Planning Authority. There are several causes for their resignation from the Committees. Most were dissatisfied because their comments and suggestions were not given importance. After serving on a Committee for a couple of years they formed the impression that it was a waste of time and had little effect on the planning decisions, so they stepped down from the Committee. Although the resigned members were completely dissatisfied with the consultation process, they said that they had high hopes not from the Committee but from the process, which could accommodate greater community concerns into the planning decisions.

There were eleven members interviewed who had resigned from the Committees for various reasons of whom some had been Committee members, and some had been Committee conveners. They identified various reasons for their resignations. None of them indicated personal reasons rather they accused the planning authority of not incorporating their recommendations, which they believed were rational and legitimate to be incorporated into the planning decisions. The following comment expresses the dissatisfaction of a former member:

> I found it [LAPAC consultation process] was a publicity mechanism for the Government, instead of an effective tool. Frankly, I think it [LAPAC] has no effect on the community. It is really a waste of efforts and a waste of time.

This former member had resigned from a Committee although he believed he was competent to be a part of the consultation process. He added:

> I believe LAPAC is poorly promoted in the community. I think most of the people in the community have very little understanding about LAPAC....People don't know what the LAPAC really does. There is a poor recognition in the community of the LAPAC and its roles and understanding, because it was intentionally designed not to be promoted enough.

Interest groups' satisfaction

The City of Canberra has over forty community and residents' groups from across the ACT concerned with environment, planning and development matters (PACTT, 2001). These associations work to develop a coordinated influence on planning and development in the ACT and are very active to keep the character of the Bush Capital and the Garden City. To maintain this character the associations also initiate some activities that are notable in the planning process. They normally meet monthly and organize discussions to address issues with other community groups.

In 1999, Planning the ACT Together (PACTT) prepared a document titled *A Strategic Plan for Canberra—Interim Community Proposal* which set out the community's needs in relation to planning and development matters. It defined the community's understanding of balanced and planned development and raised several significant concerns, including the protection of open space, areas of national heritage significance and the shortcomings of existing legislation. It also addressed related areas like transport, land release, land sale and economic issues.

The convenor of PACTT is also a convenor of a Committee. Other organizations also frequently participate in the meetings to register their concerns on specific planning issues. Since a meeting is open to all, anyone can attend and ask question with permission from the chair. Some enthusiastic members or affected people attend almost every meeting they can present their own ideas of planning issues but have no voting power. Discussed below are the views on political leaders, planning professionals and planning academics and their level of satisfaction with the current consultation process for planning decisions.

Most comments were on consultation process through Advisory Committees on planning and development matters and were directly focused on the lack of strategic plans. The views and comments of political leaders and planning professional were noted in a number of workshops on election issues organized by PACTT. Four political parties, the Australian Democrats, Liberal Party of Australia, ACT Greens Party and Australian Labor Party participated in the discussion and commented on planning and development matters. Six issues were discussed in the meeting. The issues were: (1) Strategic Planning and the City's Role as the National Capital; (2) Planning and Land Control Institutional Arrangements; (3) Community Participation; (4) Open Space, Infill and Urban Consolidation; (5) Residential Code and (6) Trees. Apart from the political parties, some planning professionals and academics were also invited separately by PACTT to evaluate the effectiveness of the current consultation process. A retired town planner was also invited to deliver his concerns on planning and consultation. His comments are systematically presented and analysed in this book to understand the breadth and depth of planning decisions from the point of view of an experienced planning professional.

A representative of the *Australian Democrats* commented on both Planning Authority and its consultation processes through Advisory Committees. He was critical of planning authority:

> PALM has been a development agency, not a planning agency. We have seen politicization of the public services. Consultation has taken place later and later. For example, Symonston Jail consultation process was at the end instead of at the beginning.

An Independent MLA also commented on the current consultation process through planning Advisory Committee and felt that comments from the community and residents were not considered by the planning authority:

> The consultation process has fallen down. The incoming government will have to take notice that LAPACs, Community Councils and Groups have their views.

The *ACT Greens Party* praised the current *ACT Consultation Protocol*. Its leader commented that the protocol was good, but not used well by the Government. This party clarified its position on the Terms of Reference of the consultation process and observed that other parties might say they would do what people wanted; however, the ACT Greens made a different comment on planning decisions and the consultation process. She added:

> Greens are not just in politics to do what the community wants. The Greens have a clear set of principles and objectives and want to work with the community to achieve the principles and objectives.

Another Greens member who was a Committee member commented on the whole planning decision and consultation process:

> Many planners have made decisions, and then just use the LAPACs as a parallel system to explain why they've made the decisions they have already made. There needs to be a big cultural change in community participation and it hasn't yet been properly taken up in the ACT.

When asked to comment on introducing an independent planning authority instead of government control, he commented that a totally independent planning authority was no good; it would still have to be linked to the ultimate authority of the Legislative Assembly. But another member of the ACT Green Party and Member of the Legislative Assembly supported an independent planning authority but expected that there must be accountability through the Legislative Assembly. This member also supported the Minister's call-in power to approve development applications when appropriate.

The Chief Minister and the Deputy Chief Minister of the ACT attended the workshop and commented on the above six election issues. The Deputy Chief Minister was in charge of Urban Services and Planning.

Since *Liberal Party* introduced PACs for community consultation about planning, they were very optimistic about getting community views for the planning decisions. However, the Chief Minister felt that there should be a change in the decision-making process to allow more people to be involved in planning. He wanted to see Committees as an inevitable part of the planning process to make more cohesive and transparent planning decisions that would ensure greater community participation. That was the reason, the Chief Minister believed, why the Liberal Government five years or so ago created the idea of Local Area PACs. However, he commented that Advisory Committees should remain as an Advisory Committee to raise community concerns in the planning decision. He added:

> LAPACs are not the supreme governors of what happens within their areas. They are an advisory process and they tell the planning process what the people in the local area think, or what a certain subset of the

people in the area think. It's a step toward more actively involving the community and it should evolve further.

The Planning Minister also acknowledged the importance of public consultation in planning decisions and commented that the consultation process should be constructive and strike a balance between local and whole:

> In principle, people who live in an area don't have exclusive ownership of the values of the area. The rest of the city has a right to take part in decisions. We have to strike a balance between territory-wide interest and the interests of the local people. LAPACs are the first step in the exercise, but obviously we have further to go.

Two frontbench members of the *Australian Labour Party* attended this workshop. Both agreed that a strategic plan was badly needed. They believed that in the absence of a strategic approach, 'planning' was reactionary rather than active and the current system reacted only to development proposals rather than adhering to a coherent strategic vision for the city. They commented that the Territory Plan was intended to be a policy document rather than a planning document and therefore needed to be supplemented.

The planning spokesperson of the Australian Labour Party argued that the best approach was to develop a series of micro-strategies for each locality, such as the 'Inner North', and for each of the town centers. These strategic plans would include social and transport planning as well as land use planning.

Although Advisory Committee plays a very important role in planning decisions, it has little influence on the final decisions. *Tony Powell*, a retired planner in the ACT who was a commissioner of the National Capital Development Commission (NCDC) back in 1975-1985, stated that Planning Authority gives only lip service to the Committee; it says their comments have been taken into account or will be given high priority in the final decision, but the agency's preconceived idea on development applications is the final one; it hardly deviates from its original propositions. Powell added that Committee's concerns are given little importance: it is the Assessment Officers who are important. Assessment Officers examine a DA going through existing planning guidelines; they use a two-column method in which selection from the Territory Plan's criteria are in the first column with check boxes in the second column to signify the extent to which the Assessment Officers consider the development applications meet the criteria in the first column. They do not look at community concerns over the development applications. Unskilled Assessment Officers with their managerial mentality address planning matters. In most cases, the Assessment Officers do not even make a site inspection, and DAs for medium to high-density proposals were fast tracked without assessment.

Powell is also concerned about the absence of skilled planners in the highest positions in Planning Authority, and thus the lack of planning understanding and of importance given to fair and effective community participation in the planning decisions. In Powell's view, the whole planning system is close to a state of collapse, especially as planning staff do not have the necessary skills to effectively communicate with the residents and community groups and to understand their expectations about planning outcomes. He also commented that the Territory Plan was considerably weakened by the Assembly under the Liberal regime in 1997-2001 and is no longer rigorous or specific enough to operate as a development controller. The objectives and principles in the Territory Plan are never applied to development assessments. The planning system borders on corrupt. Canberra has the fewest appeal rights in Australia in relation to planning issues and has moved from being the most planned city to the least planned city.

6.3 Chapter summary

The PAC comprises people from all parts of the community with the purpose of giving advice to the planning authority on various planning and development matters. This advisory group has functioned quite efficiently as a Committee, putting forth a wide variety of suggestions and concerns on planning and development matters from among the wider public in the neighborhoods. The Committee members are respected by most of the wider public for spending their valuable time to protect their neighborhoods from scattered developments. However, the outcomes and effectiveness analysis in this chapter highlights several issues faced by the Committee. Of particular importance are with its roles and operational policies in the broader context of policy planning and the planning process administered by the planning authority. All the Committee members agreed that Committee operates in an advisory capacity and provides community views on development applications. However, with respect to the effectiveness of the Committee as a good platform for public participation the members of the planning policy community are not in agreement, rather they feel it is tokenism. Some members feel that Committee has been effectively used as a link between the community and the planning authority. They also feel that Committee has great potential to be a platform for public participation but has yet to achieve this.

Some members agree that things are changing and evolving over time. Planning Authority has agreed to discuss officially in the meetings the grounds on which the community concerns and Committee advice and recommendations are accepted or rejected by the planning authority. However, despite the fact that planning staff recognize the Committee as a voluntary extension of the planning department, not a separate arm of the planning process, its primary role is to provide planning advice and not policy direction. In this respect, planning staff feel that Committee is not qualified to make broader policy recommendations and be involved at the early stages of the planning process rather it should remain community-based advisory group to voice only community concerns to be incorporated in the planning decisions.

The wider public have mixed feeling towards the Committee activities. Some are very supportive and are aware that they have residents and community representatives on the Committee. They have always hoped that Committee would be a public-oriented Committee, which would ensure a more active approach to planning. The overall feeling of the wider public is that this planning advisory group has not lived up to their expectations and they would prefer to see Committee involved in the policy process in such a way as to allow it along with the public to participate in open dialogue about planning decisions.

To summarize the evaluation of the effectiveness criteria in the consultation process of PAC, each criterion is given in Table 6.1. This book evaluated the effectiveness criteria on a three-point scale (**SM, SS** and **NS**) indicating 'satisfy most criteria', 'satisfy some criteria' and 'not satisfy the criteria'.

Table 6.1 Summary of effectiveness evaluation of consultation process of PAC

Criterion	Evaluation	Brief description
Define Role	SS	Most of the residents and community members on the Committee believe that they are competent enough to play an advisory role in consultation. However, they are not effective, compared to businesspeople who have professional ties with the planning agency. Businesspeople on the Committee believe that they are competent to play the defined role, but except the resident and community members are not.
Promote Learning	NS	The Consultation process is also a process of learning. In this case, the consultation process has no system to make all members aware of planning matters, not even basic things such the procedure of development approvals.
Communication	NS	Most of the members are dissatisfied with the communication with each other. There is no good communication either between PALM and the Committee or among the members on the Committee. Sometimes Planning Authority takes very long time to respond to member queries.
Relationship and trust	NS	The Committee appears to be divided into two groups: resident and community groups on one side and business representatives and Ministerial appointees on the other. They acknowledge that the relationship is adversarial. They have lost trust in each other.
Objective-driven	SS	The community and residents have few expectations. However, they acknowledge being often heard by the agency. They are not fully satisfied with the process but acknowledge some improvement on the issues they mentioned for rectification.

Notes

1. Kingstonization is derived from the suburb of Kingston, which has had high pressure of multi-unit development in recent times. Members use the term 'Kingstonization' for more multi-unit development in any suburb.
2. B13 is a part of land use zoning in ACT. There are three types of land use zoning for residential and commercial development: B11 allows two storey, B12 three storey and B13 nine storey buildings (PALM, 2000a).

7 Interpretation and discussion of the findings

7.1 Introduction

Chapters 5 and 6 evaluated the proposed criteria of fairness and effectiveness of the participation process through Local Area Planning Advisory Committee (PAC). The criteria were evaluated by critically examining the context and its operational processes, and conclusions were drawn after discussion of significant issues identified by PAC members and others involved in planning decisions. Although Committee members expected that through the consultation process their advice and recommendations would be considered by the planning agency as community input, they often believed this did not happen. Residents and community representatives on the Committee particularly felt that they were not influential in the planning decisions and that it was developers who had the most influence.

The purpose of this chapter is to discuss these negative implications of the consultation process. Specifically, this chapter presents a critical analysis of Committee's approach to public participation by examining key issues involving Committe members and others in the planning policy, and show how the existing context and process of a PAC may be improved. The chapter also identifies the barriers to fairness and effectiveness of the participation process in PACs. The analysis is divided in two categories: (1) differences in perceived understanding of fairness and effectiveness in the consultation process and (2) identification of the barriers to fairness and effectiveness.

The following discussion expands beyond the analysis of the interviews and encompasses information provided in the previous two chapters. It both interprets and discusses the criteria for evaluating effectiveness and fairness using interviews, observations and documentary data to understand the perceptions of the individuals involved in consultation processes.

7.2 Differences in perceived understanding of fairness and effectiveness

PAC is a committee of diverse people comprising residents, community groups, businesspeople and Ministerial appointees. It includes people with

DOI: 10.4324/9781003122111-7

professional and non-professional planning expertise. Non-professional planning experts are those who have long been involved with the consultation process and become proficient in planning matters. The perceptions of the varied groups about the consultation process and its outcomes are different and conflicting. They mainly indicate the negative aspects of the current consultation process, which requires immediate attention to make it fair and effective. Therefore, this section highlights the perceived negative aspects of consultation process and the expectations of the planning policy community.

7.2.1 *The limited opportunity for learning*

Involvement in the participation process has always been an opportunity for the participants to learn more about it. While participants may be expected to learn a great deal about the details of planning decision-making from their continuing activities, the learning process has not been successful in the case of PAC. The criterion *promote learning* is especially important when a Committee comprises of people of diverse backgrounds and shares the information with each other to contribute effectively to the outcomes of the consultation. The Committee consultation processes lack the opportunity to educate the participants on planning and development matters because there is no opportunity for such education within the Planning Authority administrative process.

New members on the Committee expect to be adequately briefed about the context, operational process and results of planning decisions. Particularly, what is needed at the beginning of the consultation process is a working knowledge of planning laws, the approval process, the design guidelines with which development proponents should comply and familiarity with associated planning information such as the Territory Plan and Guidelines for achieving High Quality Sustainable Design (HQSD), and the administrative process of development approval. Other information related to development applications is also important in making useful comments. However, it was noticed that the new members were not adequately briefed on planning laws and design guidelines, particularly the process of development approval, although understanding of these issues is fundamental if the members are to comment on development applications. A Committee member, by profession a planner, asserted that the planning authority was not sufficiently committed to organizing workshops or orientation programs for members to help them understand planning and design matters. Without such programs, new members are not competent to make comments on development applications. A business representative said that:

> Having served for over five years or so, a member asks what does the Y-Plan mean? Is it Territory Plan? How do you expect informed comments...on HQSD stuff?

However, there are different views among others involved with planning decisions. Some resident members believe that the Planning Authority should help the Committee members to adequately understand the decision-making process and, most importantly the process of development approval; but at the same time, members should be interested enough to learn about the relevant planning laws and design guidelines in order to make competent comments.

It is noteworthy that Planning Authority does not organize any educational activities for the Committee members. It appears that planning understanding is the sole responsibility of Committee members and that Planning Authority can do nothing to educate members to understand planning issues and related matters. But planning understanding should not be left to the Committee members. The Planning Authority has a major responsibility to organize outreach programs to provide an organized process that members should follow throughout their tenure. In this context, Webler (1992) comments that a competent consultation process is not fully an individual task or personal matter, rather it is procedural competence, which ensures adequate access to the information and easy-to-understand planning documents and helps discourse participants to make informed comments on planning initiatives. Planning Authority does not do this, so the new members take a long time to become familiar with the consultation process, particularly the process, and guidelines of development approval.

7.2.2 *Lack of adequate training and communication skills*

Long-time members of the committee are always critical of the planning staff. They feel that planners need more training to improve their communication skills. Some non-professional planning experts on Committee raise concerns about the skill and efficiency of planners in communicating with others involved in planning policy, particularly with the Committee members. They believe that planners need training on how to conduct effective consultations with PACs. They further believe that the planners need skills in arranging planning workshops, understanding the social values of residents regarding planning decisions and competently explaining planning laws and design guidelines. Some experienced members on the Committee feel that planners have only temporary obligations to attend the meeting and often show no serious commitment to community concerns. Some tend to skip the members' concerns on the planning issues, and often, they cannot provide necessary planning information to the members. This is not the case in all Committees, though. Planning Authority usually nominates experienced planners to some Committees that experience high pressure for development in their areas, caused by many development applications to be approved in a limited time. Manuka and Inner North Committees are the examples of representing Planning Authority by experienced planners, but the other four Committees have no specific planners to represent the Planning Authority.

Six planners were interviewed and found none had received any formal training in conducting the public participation process with a regular PAC. All of them said that they had adequate formal training on conducting planning workshops on community need assessments and urban design projects, but they did not feel special training was necessary for public consultation with PACs. They believed that skills in public participation were innate and could not be acquired through training; instead, skills could be learned through trial and error. Within the Planning Authority administrative systems, there are currently no training and outreach facilities to guide the planners in the development of public participation strategies and in effectively conducting the consultation process. There is a consultation manual prepared by the ACT Chief Minister's Department entitled *Consultation Manual for the ACT*, which gives advice on both statutory and non-statutory consultations, but few planners seemed to be aware of this consultation manual.

Comments made by the planners indicated their lack of understanding of the importance and complexity of conducting the public participation process. However, without adequate communication skills, planners cannot meet their responsibilities to understand community expectations. Kaufman (1999) pointed out that the root cause of inadequate communication skills and lack of understanding of the community values was due to the lack of adequate planning education in Australia, which meant that social and community values could not be incorporated into the planning decisions. She also observed that planning education in Australia did not give planners sociological insights, which prevented them from adequately understanding community values and communicating them effectively. In Australia, a planning degree requires five years that include thirty units of which only one sociology unit is required. So, planners often misunderstand community values and social preferences.

Similarly, a member of the ACT Planning and Land Council (an expert advisory committee on planning and development matters, now called National Capital Design Review Panel), Brendan Gleeson and a planning academic Nicholas Low state that Australia's planning education lacks a proper planning course to train its students to conduct effective public participation programs and to communicate with the planning community (Gleeson and Low, 2000). They proposed that public participation courses should be introduced at both the undergraduate and graduate levels (Gleeson and Low, 2000: 232). Self (1998) perceives the present planning direction as 'Market-driven' and playing down community values in planning decisions. He commented that the contribution of professional planners is downgraded, and planning decisions are increasingly made based on developers' pressure or political opportunism rather than professional analysis. This appears to be the case in the ACT's consultation process. A retired Town Planner commented that the ACT had started de-skilling planners in planning decisions and that when the ACT became self-governing, planners began leaving it: re-skilling is needed to address planning and development matters.

7.2.3 Lack of adequate competence understanding about the process

Some business representatives raise questions about the competence of Committee members and their understanding of the planning process. The interviews with business and Ministerial appointees indicate that they are critical of the resident and community representatives. They believe that residents have no adequate understanding of planning and cannot articulate their concerns effectively before the development proponents, rather, all they have is emotional attachment to their home areas. Their comments on development applications are not rational arguments worth considering, and do not present relevant facts or justify due attention. In response to this opinion posed by the business representatives and Ministerial appointees, a resident member on the Committee commented:

> LAPAC members could not be seen as experts, but as 'non-professional planning experts'…involved in the consultation process for a long time and well aware of its process. We do not have to be professionals to be included on planning committees. We are part of the community only.

When this respondent was asked to respond to comments such as 'Most of the Committee members have no competence level of planning understanding" and "Some cannot point to the north point of a map', he replied:

> Our job is not to be expert in reading plans and maps. The committee's concern is to understand the impacts [of development applications] on their localities, and LAPAC is not a technical advisory committee, it is a PAC… consisting of people with or without planning understanding. What we need to know is the planning guidelines and decision process.

7.2.4 Lack of social planners to capture social values of the Committee members

The absence of social planners in the Planning Authority is intensely felt by the members, and it is known that social planners are needed in conducting consultation process through a PAC. Currently, the Planning Authority is dominated by planners with a background in architecture and landscape architecture. There are no social planners for any statutory consultation processes, and some long-time Committee members feel that their absence prevents from capturing the community's social values and feelings of attachment to the places where they live. A resident member cited a proposed development application known as DV-200, (also known as 'Garden City Variation') which will significantly change the character of Canberra and will lift some existing restrictions on residential and commercial development in the inner-city suburbs. Resident and community representatives on the Committee comment that the value and recognition of the national

capital appear to be of little importance to the Planning Authority and development proponents when they propose to change many characteristics of the
'Garden City'. The Committee members felt that residents and community
views on the DV-200 had not been considered because no importance had
been given to social and environmental values. There are no social planners
to encapsulate the social realms and 'sense of place' of the residents, neither the planning agency nor the developers give importance to the social
implications of development applications. They comply with planning and
design guidelines only and do not adequately explain and discuss the possible
impacts of development on neighborhoods. A member of the wider public,
who regularly attends the meetings commented:

> Neither PALM nor developers have social planners. All they have are
> some architects who cannot, I believe, carry out community consul-
> tation effectively that would reflect community values, and the social
> fabric.

Another member of the Committee pointed out that the community could
not influence the planning decisions as they were not given adequate time
to make comments and recommendations: only three weeks in the case of
'DV-200'. This advisory group felt that the allocation of time for commenting on 'DV-200' was not adequate for them nor for the wider public. Other
members on the Committee noted that they lacked any significant influence
on planning decisions. A retired planner commented:

> People or any consultation body in ACT have no real influence on the
> planning decisions. LAPAC, as a consultation body, was working well
> and raising concerns that were best suitable for their own localities, but it
> could not effectively influence to the better planning outcomes that were
> suitable, and based on regional goals to foster Canberra as a liveable city.

Powell added that planning decisions were based on a political agenda. The
agenda was to sell and develop or to redevelop land if the developers wanted,
and Planning Authority acted to serve political decisions only, because
Authority lacked skilled planners that could foster the social values of the
community and enhance the importance of Canberra as the nation's capital.
In effect, Advisory Committee and other planning organizations were only
influential in delaying the approval process. However, at the end, the developers would win although they might agree to some minor things to their
development applications.

Powell suggested that the consultation Committee should have competence enough to put forward the views, which were believed to be acceptable
to the greater community and to the Planning Authority for consideration. For him, in order to achieve this goal, all consultation Committees
should include academics, community leaders and non-professional planning

experts. However, he was sceptical of the likelihood of adequate budget and secretarial support for the Committees to collect community input on development applications.

7.2.5 Lack of amicable relationship among the planning policy community

Overall, the consultation process through the Advisory Committee indicates poor relationships among the planning policy community and distrust among them. Knaap et al. (1998) point out that consensus is a pre-condition for trust among the discourse participants, focusing on shared objectives. However, there is little sign of information being shared between business representatives on one hand, and residents and community groups on the other. It was noticed that the relationship between business and community groups was always antagonistic, but there was a good and trustworthy relationship between residents and community groups. Such trustworthy relationship existed due to a common vision on the development and redevelopment of their suburbs, a strong commitment to protecting them from what they termed as the 'greedy mentality' of development proponents, and a common and shared love of the places where they had been living for a long time.

Conflicting views

Conflicting views on development and redevelopment issues among Committee members, contributed to a loss of local character and identity. The business representatives, i.e., the *de facto* representatives of the developers, wanted more redevelopment and affordable housing units around the inner city. They said that there was a high demand for affordable housing, particularly single units, in those areas, for middle-class single employees working in and around the city. But residents and community groups wanted to preserve the character of the 'Garden City' in their suburbs by restricting multi-unit developments and prevent permission for dual occupancy on big blocks. These conflicting opinions on urban renewal among residents and business representatives may lead to many changes in the urban character. The resident members believe that business representatives are always found on the Government side to support development applications. A resident member commented:

> Business representatives on the LAPAC are government agent to approve DA, without much obstruction.

The most obvious conflict among the members was observed in the Manuka and Inner North Committees, each of which appeared to be divided into two main groups: the residents and community groups on one side, and business and Ministerial appointees on the other side – a perspective of *us* and *them*. A resident representative on the Committee commented:

> Business representatives want to bulldoze all the trees and one or two-storey buildings, and put ten-storey buildings in Northbourne Avenue, which is ridiculous development in the gateway to the national capital.

There was conflict over the allocation of local open space for development work. The community felt that local open space would be gradually taken up by putting units around the open space, the rest of which would be taken up subsequently near the units proposed for development applications. This was exemplified by a proposed multi-unit development along the Barry Drive in Turner. The community felt that this development would jeopardize open space for the residents living in the neighborhood and remove pedestrian access for the commuters going to the CSIRO and ANU, but they had not been successful in keeping this area as open space. The flexibility of the Territory Plan allowed developers for variation to the Territory Plan to accommodate various land uses. The resident member commented:

> Decisions have already been taken for development of this park. Where is the consultation? Here consultation means to make comments on decided proposals. When necessary they [planning authority] will make a little change to prove that community concerns are being taken into account. It is like window dressing. It is neither participation, nor consultation, it is just tokenism.

However, business representatives have different views. They believe that responses from the community are normally given high importance by the developers as well as the Planning Authority. They believe that Canberrans are 'over-consulted'. In every planning initiative, Planning Authority conducts consultations with both statutory and non-statutory groups. A business representative said that people should be happy as they were being informed about what would happen in future; they had adequate opportunity to their concerns during the meetings and later could send a submission detailing their concerns on proposals, which would eventually help Authority to make a balanced planning decision. A Ministerial appointee commented that most of the community's and residents' views on development applications are based on the 'NIMBY' (Not in my backyard) syndrome:

> They [LAPAC resident and community groups] don't want some units around them. Other income groups have the right to live near their workplaces. As a developer [here as a business representative in LAPAC], we are meeting housing demands. We are not putting bricks; we are concerned about high quality sustainable development as well.

Similarly, the Kingston Foreshore Development Authority, in its first phase of work, would provide 1865 dwellings just behind Wentworth Avenue in Kingston. The concern of the community was the 'scatter-gun' approach to planning where developers initiate development proposals, which sometimes need variation to the Territory Plan. Ultimately, continuous variations to the Territory Plan, the community believes, would destroy the natural environment of Lake Burley Griffin. Resident and community groups have had also been critical that the role of National Capital Authority has not been effective in introducing development control plans, which the community believes need to be stronger to preserve the elements of the National Capital and Federal Parliament.

Kingston is under the Manuka Committee. The residents and community representatives believe that most of the business representatives on this Committee have business relations with the Kingston Foreshore Development Authority, and always support to what the KFDA proposes to do. A resident member said:

> Business representatives are here as if they wanted to see that all development applications from the Foreshore Authority are approved, which will probably give them some extra business. The community need is nothing to them.

These differences of expectation and conflicting views of the role of the Committee in the consultation process may give a relative autonomy to the Planning Authority to implement their own planning options and undermine community preferences. Webler (1995) commented that any fragmentation among the discourse participants gives relative advantages to the decision-makers, which often undermine overall community concerns and their preferences for development.

7.2.6 Lack of opportunity for early involvement

The lack of opportunity to participate in the early stage of development applications to make informal comments means that, often the members do not fully understand the extent of a proposed development application and what it affects, until development work has commenced. Many members, particularly in the Inner North and Manuka, have identified the lack of early involvement as a major problem of development applications. They believe that DAs are put on the meeting agenda after many things have already been done, including formal and informal consultations with Planning Authority and other planning sections. Residents and the community are often not consulted before lodging formal applications. A Manuka Committee member commented:

> Before putting a formal DA with PALM, they [developers] start bulldozing trees prior to consulting LAPAC and the communities, as if their DA would be approved anyway.

Another Committee member agreed. He was also critical of the huge rede-velopment along Northbourne Avenue and multi-unit development in big blocks in O'Connor, adjacent to Northbourne Avenue. He believed that developers could hide many things when they start construction:

> They [developers] can easily hide by putting a wall on the construction side. They demolish trees when they feel necessary. Nobody can realize what is happening inside. If you ask why they demolished the trees, the reply was to plant ornamental woods in a suitable position.

7.2.7 *Lack of adequate time to review development applications*

There is a widespread complaint that Committee members are not given ade-quate time to review development applications. They feel that some devel-opment applications have national interests and need greater consultations before final decisions are made. However, the general feeling is that people must ask Planning Authority in almost every development application for a greater extension of time. This would allow the wider community and other interested individuals and groups, time for an informed submission. This occurred with the DV 200 proposal.

In this regard, most of the Committee members identified two impor-tant reasons why Committees were not given importance in the consultation process. First, Planning Authority and the development proponents did not whole-heartedly feel that community input is important in planning deci-sions. Second, they did not give real value to Canberra as the nation's capital, which has a unique character to be preserved; consequently, they also failed to value Canberra's international reputation as a planned city. A member commented:

> If you believe in the Garden City, Planned City, and Bush Capital con-cept, you must preserve its uniqueness. But politicians are devaluating its international image. And proponents [of development] want to make money devaluing its national significance.

7.2.8 *Lack of proper representation on the Committee*

There is a problem with representation on the Advisory Committee. The Committee consists of residents, community groups, businesspeople and Ministerial appointees. The residents and community groups are dissatis-fied with the business representation on every Committee. But the press-ing concern is: Who will be on the Committee? How can it ensure greater representation from all kinds of people? How can it get the whole com-munity's input in the consultation? So, choosing the membership for the Advisory Committee is a critical issue. Also, it seems that only known activ-ists, and community and political leaders are normally chosen to serve on the Committee in the hope that their recommendations and advice will be respected by the wider community. However, the community interest may be spread and not well organized, and therefore, not adequately represented on the Committee, so the whole process may be disrupted by challenging the legitimacy of the Committee (Webler and Tuler, 2000). The Committee representation is no exception so the wider community raised this fundamental question about representation, which they termed a 'small segment of hand-picked elites'. As such, the Committee members believe that they do not have adequate opportunity to collect community input and convey it to the planning agency for consideration.

The *election and selection process* for Committee membership has been criticized by both the wider community and the Committee members. They say that the election process is fair, but the concern is: who is going to be elected finally? Some members said that inadequate profiles of Committee's activities do not encourage the wider community to be part of the consultation process. There is also criticism of the protocol of Advisory Committee because there are only residents, community groups, business groups and Ministerial appointees on the committee. Other representatives of the community such as schools, churches, sports, police, housing, property management, youth and disabled people are not on the Committee.

There is also concern about outsiders on the Committee. Members who are neither residents nor in business in the neighborhood may not clearly understand the needs of residents and the community at large. Webler (1995: 39) commented:

> ... membership of the advisory committee is typically chosen from among the leaders of the community. They belong to the same class of elites as the government officials, experts, and stakeholders. They are more likely to rely on instrumental understanding of the problem and downplay the value of anecdotal evidence and competing normative arguments.

Therefore, Advisory Committees should be formed in way that maintains a balance of representation and people who have genuine interest in the area and its planning outcomes because a balanced membership is essential to establish trust and reduce conflict. Some members of the Committee feel that Committee itself contains no specific guidelines regarding its balance and that, residents, community leaders and some selected Ministerial appointees do not represent the whole community. The Planning Authority is responsible for interpreting what constitutes balanced representation. Long and Beierle (1999) note that the criteria for balanced membership in Advisory Committees may vary according to the subject matter of a group. Since Advisory Committee discusses planning and development matters, ranging from the hydrological aspects to the social implications of development applications, some members feel that there should be some technical people on the

Committee to qualify the comments on very technical aspects of the proposals. According to Long and Beierle (1999: 11):

> Thus, where the issues are broad and policy-oriented, advisory committees should also be broadly representative, along technical, social and political dimensions.

A planning staff member commented on representation on advisory committee:

> LAPAC does not truly represent the community, as it has only members of retired persons, community leaders, long-time residents and business-people. It has no wider community representation on the committee such as youth organizations, church leaders, single mothers, and disabled people.

Most of the planners also acknowledge that residents, community groups and business interests alone do not truly represent the whole community on the Committee but say there is other representation as necessary. Developers have no representation on the Committee, but business representatives work on their behalf because most of the business representatives are developers. The Committee protocol should clearly define the businesspeople and developers on the Committee. Since developers play a very important role in the meeting, presenting applications for development, they often insist to Committee members that as the applications comply with the guidelines, therefore, they should be approved, thus, reducing the opportunity for members to express necessary concerns about them. A Manuka Committee member said:

> Often, we have been threatened by the developers that they comply with the Territory Plan, and we have to accept it. Because, the Territory Plan allows them to do it. There is no other avenue to find alternatives, as developers believe that they comply with planning rules and regulations and also comply with Design Review Panel (DRP) guidelines. We feel disenfranchised.

7.2.9 Inadequate and limited role

Even though some long-time resident members want to have a combination of professional and non-professional planning experts on the Committee, overall the Committee is neither an 'Expert Advisory Committee' nor a 'Policy-level Advisory Committee'. Planning Authority acknowledges that the Committee is simply a 'Tasks Specific Advisory Committee', representing diverse community interests and providing a public forum for members of the community to attend and discuss their needs and concerns about the decision-making process. All Advisory Committees make comment only on development applications. They were formed only for this purpose and not to make comments at the level of strategies. It appears that Committee's function is to comment on development applications, which have been submitted by the developers with or without prior consultation with the Committee and other community and residents' associations. However, DRP appears to be an expert Advisory Committee, consisting of planners and academics, whose comments, the Committee members believe, are given high priority.

DRP discusses planning and design matters. This panel has generated distrust and disillusionment among the Committee members as it also reviews all development applications in order to examine whether they comply with Territory Plan and design guidelines. It appears that Committee and DRP may have competitive roles in the consultation process. Observation reveals that it is worthwhile to avoid parallel bodies discussing the same issue because their introduction may generate greater distrust among other key planning stakeholders and thus increase conflict and cynicism.

Two DRP members who were officially Technical Officers were interviewed. They explained that DRP was not a parallel organization but provided planning and design guidelines to the development proponents for achieving high-quality sustainable development. But the Committee members interviewed have the impression that DRP was purposely created to undermine the Committee's suggestions and to rely on expert opinion. The wider community believes that this type of parallel organization can have more influence on planning decisions than community orientated statutory bodies. A Committee member commented:

> We are a task-oriented group. If they [PALM] feel our comments have value then they will give them importance, otherwise, they will ignore them.

He added:

> Why are we a group, since we have been encouraged for only individual submissions, not a body to submit any concerns collectively? It is carefully designed not to be very effective in planning decisions in order to ensure that the planning professionals and proponents can implement their own pet hobby-horse ideas.

7.3 Barriers to fair and effective participation

The evaluation of Committee's role has revealed two important aspects of the barriers for fair and effective participation: first, differences in understanding over the role of the Committees in planning and development matters, and second, value-driven expectations by the Committee members about the planning outcomes.

The Committee protocol outlines a very specific role for the Committee, which is to advise the Planning Authority on planning and development matters referred to them. However, not all planning and development matters are included in the Committee's agenda. As discussed earlier, some planning issues are excluded, such as dual and triple occupancy in a big block, transport planning, land sales, land release, areas under commonwealth control and park and open-space development and management. The exclusion of some planning issues from the Committee may discourage members from playing effective role in overall planning issues. Renn et al. (1995) argues that public participation process is unfair if participants are not able to discuss all related issues during the meeting. The explanation of exclusion of these planning matters in the Committee's consultation process are also not acceptable to the members, who believe that government has come to a wrong conclusion that the Committee is overwhelmed with DAs. This is not a justifiable explanation.

The Committee's primary role is to act in an advisory capacity, but interview data indicate some disagreements on whether the Committee should be simply advisory or as a planning watchdog. These differences in understanding the role of the Committee may result in conflict and arguments during the meeting. The members acknowledge that sometimes Committees have difficulty in putting forward their recommendations collectively or initiating their own activities jointly when there are discussions about the clarification of their role in planning decisions. Resident members felt that Committee should be a planning watchdog to convey their needs and expected planning outcomes to the Planning Authority, but some business and Ministerial appointees believe that the Committee should remain only as an advisory body.

These different views have also been reflected in Planning Authority's planners, who hold very different views about the role of Advisory Committee. It is felt that planners see Committee as part of their consultation obligation to facilitate decision-making. The Committee is a group to which planning staff can give information and planning documents for comment, not for generating new planning and design ideas. The Committee should be a PAC to obtain comments and feedback on development application on behalf of the community. The Planning authority will value their advice and opinions in the planning processes, but not necessarily give them the highest priority. The Committee should remain a public forum for residents, community activists and businesspeople, but not as a forum for policy planning. The policy planning should remain with the government and its Planning Authority. The planning Minister said that the current government is community-oriented and would give value to community voices; however, he felt that decision-making power should remain with the Legislative Assembly. But decisions would not go beyond the community's feelings and the input of the population would be given high priority in the final decision-making.

The Planning Authority seems to expect the Committee to act as a means of public participation, where development proponents, the wider public, and enthusiastic planning groups can bring planning and development concerns for discussion. The Planning Authority also seems to expect that once development applications are tabled; the Committee may make recommendations regarding their preferred options. But these recommendations do not receive any response or feedback except for acknowledgement of them. In this way, the planners can filter Committee's concerns and apply their own judgment on development applications.

But, the views of the wider public are also pertinent to the consultation process. Members of the wider public perceive Committee's role as a source of information or fact-finding body for the Planning Authority. They believe that the role is limited if it can only act in an advisory capacity to Planning Authority and the Minister for Planning. They believe that the Committee provides a platform for the wider community that might not otherwise exist – they can write letters or appear at meetings as a way of introducing their concerns.

These differences in perception of the role of the Committee indicate some serious barriers to the fair and effective operation of the Committee's consultation process. Community groups believe that the Committee should be a watchdog, while businesspeople, Ministerial appointees and planners believe that it should merely report community concerns. The wider public believe that the Committee is a form of public consultation. It seems that planners and Planning Authority use the Committee to review public concerns on planning issues, while the wider public use the Committee as a good avenue to formally register community concerns to possibly affect the planning decisions. It is evident there are different perceptions of the role of the Committee and conflicting expectations about the planning outcomes. So, it seems the Committee is not successful in fulfilling its role.

7.3.1 Response and feedback

Responsiveness and timely feedback to the members greatly influence the perception that their input and recommendations have been considered in making final decisions, but Committee members complain the lack of such responses. Planners often do not provide feedback. And sometimes, one planner is replaced by another at the next meetings. Such substitute does not know what happened in last meeting and cannot provide feedback to the Committee.

Committee members were often told that their questions had been forwarded to the relevant section within Planning Authority or the ACT Government and were resentful when they did not receive a reply. This lack of feedback and accountability to the community may lead to the assumption that community concerns are not given importance or taken into consideration, which causes some long-time members to have little enthusiasm for remaining on the Committee. Most of the former members who had resigned complained about inadequate feedback to their concerns about whether they were heard.

7.3.2 *Influence and power of decision-making in planning*

The power of decision-making in planning and development matters rests mainly with the ACT Government and its Planning Authority, but the main political decisions are in the hands of the Minister for Planning, who has call-in powers to undertake final decisions.

However, Planning Authority sometimes makes it own decisions without prior consultation with the Minister regarding non-master-planned development applications, which only require a very small change in existing land use. But when the development applications need greater public involvement and create disagreement, the Minister uses call-in power to make the final decision. A Committee member commented:

> We don't know why the Minister uses the call-in power. There is nothing written on the Minister's call in power, on what grounds the Minister uses it. Only the LAPAC protocol indicates that Minister has call-in power. But there should be a written statement on what grounds the Minister uses it. There is no document and written statement detailing call-in power. It is like a communist state only with gentle behaviour.

There seems to be 'top-down' approach to planning-decision practices in the ACT. This approach is mostly consultative and dominated by the comments of the planning professionals of Planning Authority and the developers. Most members feel they are always given opportunities to raise their concerns over development activities, but they also feel that planning professionals have the most influence on decisions, if there are no political or election commitments by the Minister. The planners who examine the proposed plans have commented that the consultation approach is 'top-down' and they are trying to bring a 'bottom-up' level to give more voices to the community and residents in planning decisions. A planner added:

> Those who live, play, work, and invest in their areas have to be the main planning partners in neighbourhoods. The current process gives little opportunity to the community and residents to have their say incorporated significantly into the planning decisions. The neighbourhood plan is working to bring the community and residents to be equal partners for planning decisions.

However, the Planning Authority is very close to the community but it has very little power in decision-making, particularly when political decision and approval at government level is required. In a broader sense planning goals are sometimes tied to the political and economic concerns: examples are the Kingston Foreshore Development and multi-unit housing development along Northbourne Avenue. The Planning Authority has nothing to do with the inner-city housing development close to the Lake Burley Griffin but gives only professional advice and support to the whole process. Most of the members expressed concerns that these are political decisions and the process has been initiated by the development proponents with planning staff as a partner. Similarly, the proposed ACT Jail in Symonston generated huge objections from the residents and community groups, particularly in the nearby suburbs of Narrabundah, Griffith, Forrest and Red Hill. Their concern is that its proximity to suburban areas would have a great major impact on the locality and its landscape. This view has been acknowledged by planning professionals, but they commented that it was a political decision to select the right place to construct a jail for the ACT. A developer expressed different views:

> Relatives of prisoners have the right to see prisoners within close proximity of their residence. They [prisoners] are not boat people to be sent to Woomera detention centre [a detention centre for illegal migrants]. The ACT Jail is long overdue, and it is necessary, but the question is that everyone will say "not in my backyard".

7.3.3 Consultation with known stakeholders and with decided proposals

The Committee members feel that the community is consulted mostly on already decided development activities. The comments of community groups and long-time residents on Committees suggest that the consultation process is based on what Webler (1992) termed the *decide-announce-defend* approach. In this approach, the Planning Authority first *decides* with the development proponents what to do and later they circulate (*announce*) plans in the form of either development applications or Master Plans among the planning stakeholders. In the name of consultation, the planning authority simply *defends* what they have decided to do. The statutory consultation with the advisory groups and other planning stakeholders provides an opportunity for community input that is often limited by inadequate provision of information to members on development proposals. Such consultation process does not give adequate response to individual members or Committee, gives a limited time in which to respond and lack of formal feedback to all Committees as indicating whether their comments are being considered into the planning decisions.

Since 1995, Planning Authority attempted to involve community-based planning organizations such as PACTT, Save the Ridge, Community Councils (North and South) and Friends of Aranda in the consultation process. However, most of the development applications were lodged before the planning stakeholders were timely consulted. Residents and community members said that Planning Authority normally consulted planning stakeholders on development applications, but most of the time the applications were submitted before the consultation process starts. A member commented:

We see a DA when it has already started working, we have been given so many options, but they don't invite us to create options, instead they give us options to choose. As if the initial process for generating options is the sole responsibility of the planners. We are given no chance to make our own options, but to accept their options, which they designed with prior consultation with the concerned [development] proponents.

Accordingly, Planning Authority conducts non-statutory consultation events with many groups such as high school students, community groups, residents' associations and possibly affected communities and individuals. Consultation by Planning Authority on the Jamison Group Centre Master Plan was with Canberra High School students and caregivers at the Aranda Nursing Home, as the school and nursing home are close to the Jamison Center. Most of the members commented that such processes were just whitewashing to reduce conflict and persuade the public into feeling that they had a voice in planning decisions. However, a planning staff member commented that this move might be used as a model for further consultation with various groups such as youth organization, single mothers, nursing home staff, social clubs residents' groups, religious leaders, ACT Housing, the Defence Housing Authority and environmental groups, regarding major development or redevelopment work in their areas.

7.3.4 Barriers to fair and effective participation in Planning Advisory Committee

The planning stakeholders identified the following examples of barriers to fairness and effectiveness in the consultation process. Observations by attending the meetings and other planning documents are also included in the lists of barriers. From the identification of barriers to participation in the PAC discussed in Table 7.1, it appears that Committees do not adequately provide effective planning input for strategic planning initiatives. This is not only due to different expectations of planning outcomes by different actors the planning policy community, but also an indication of problems in the Planning Authority. The planning policymaking is an interactive process and requires collaboration among all in planning policy community to ensure a fair and effective consultation process. The planning policy community should feel that the process is carried out fairly and effectively, and that committee members have adequate opportunities to influence planning decisions. However, major planning stakeholders believe that most of the planning decisions take place behind closed doors in consultation with development proponents. The residents believe that planning decisions are largely influenced by planning officers and are based on their recommendations; the wider community has similar views about the effectiveness of Committees in the planning decisions. The analysis and discussion in Chapters 4 and 5 indicated the barriers

Table 7.1 Barriers to fair and effective participation in PAC

Barriers to fair participation	Barriers to effective participation
Adequate opportunity	*Defined role*
Inadequate time to review DAs	Limited role on planning matters
Inadequate time to lodge submissions	Only comments on DAs
Inadequate access to planning documents	
Inadequate explanation in documents	
Lack of adequate preparation on DAs	
Lack of opportunity to lodge complaints with AAT	
Early involvement	*Promote learning*
Lack of early involvement in the process	Lack of willingness to learn
Too late to receive planning information	Lack of language accessible documents
Not involved in initial planning stage	Lack of ability to understand needs
	Lack of planning understanding
	Lack of ability to challenge
	Not updated with current planning laws
	Inability to articulate planning matters
	Inability to capture social values
Feedback and responsiveness	*Communication*
Lack of accountability of the planning authority	Lack of effective communication among planning community
Inadequate explanation in feedback	Lack of effective communication between DA presenters and members
Lack of provision to give feedback in protocol	Absence of information technology for communication
Inadequate feedback from development proponents	Lack of interpersonal communication to share planning information
Lon time to respond to committee concerns by the planning authority	Lack of effective communication between planners and committee members
Receive feedback that was not suggested	
Members do not know where the input has gone	
Agenda setting and minutes-taking	*Relationship*
Prevailing hidden agenda	Adversarial relation between residents and business representatives
Exclusion of planning issues from the committee agenda	Hiding from each other
Lack of accurate minute taking	Lack of consideration and compassion
Committee advice often does not reach to the Minister	Lack of trust among planning policy community
Planners can filter committee advice in minutes	Lack of tolerance to accept other people's assertions on planning matters
	Prevailing opponent mentality
	Lack of willingness to establish friendly relations with planning community
Representation	*Objective-driven*
Lack of wider range of representation	Lack of available evaluation of existing process, if modification of practices is required
Lack of knowledgeable individuals	
Inappropriate selection criteria	Lack of evaluation of members' expectation
Inadequate profile of committee	Lack of evaluation to understand whether members' expectations met
Excessive use of call-in power to nominate members on the committee	

to fair and effective consultation through a PAC. The elements that need to be addressed for a fair and effective PAC are discussed in the next chapter.

7.4 Chapter summary

Desired expectations about planning outcomes among all the planning policy community vary significantly and are often conflicting. The book indicates that the residents and community groups are on one side and the business representatives and Ministerial appointees are on the other. The consultation occurs on urban changes, which may result in the loss of Garden City elements, local character and identity, with bad effects on long-time residents. Therefore, residents and community groups have established NGOs outside the formal planning process to cope with such effects and raise planning concerns to be addressed by the Planning Authority.

The Committee members expressed concerns for a fair and effective consultation process. They want a process, which is more transparent, open and accountable to the residents and the community. Planners reviewing different planning proposals acknowledge that the consultation process in the ACT inherited a traditional form of 'top-down' approach, while a 'bottom-up' process needs to be promoted. Chapter 8 addresses issues that are essential to re-structure the consultation process to develop a conceptual model for a PAC that will work in a fair and effective way.

8 Conclusions

A conceptual model of fair and effective public participation process in urban planning

8.1 Introduction

The public participation process is too complicated to come to a definitive conclusion on what constitutes a fair and effective process. This book has envisioned addressing this question through evaluating consultation processes of Planning Advisory Committee (PAC) based on the concepts of fairness and effectiveness. The qualitative analysis described in previous chapters and the analysis of findings described below lead to develop a conceptual model of the public participation process for the PAC. This study also provides guiding criteria and question sets for evaluating fairness and effectiveness in PAC that may apply in a more general sense.

This book adapted evaluative criteria from relevant literature (Webler, 1992, 1995; Beierle, 1998; Octeau, 1999; Chess, 2000; Rowe and Frewer, 2000; Webler and Tuler, 2000); however, the focus was mainly on evaluating PACs as a method of public participation process. This book has examined criteria that will be suitable and appropriate for analyzing the PAC and its consultative process. The evaluation was carried out through attending meetings, interviewing the planning policy community and analyzing available documents on the consultation process of PACs.

8.2 Conceptual model

The goal of the conceptual model (Figure 8.1) is to produce a process that is fair and effective and to recommend evaluative criteria and question sets for the PAC. Chapter 7 discussed barriers to the consultation process and consequently the question arises how the Planning Authority would be able to introduce notions of fairness and effectiveness in participation process. A number of conceptual models have emerged in the literature on environmental decision-making to address such fairness and effectiveness, but there are few literature sources that evaluate criteria of fairness and effectiveness of public participation process in urban planning, particularly in cities where redevelopment pressure is very high. Since Canberra is known as a planned city, has international reputation as a Bush Capital,

DOI: 10.4324/9781003122111-8

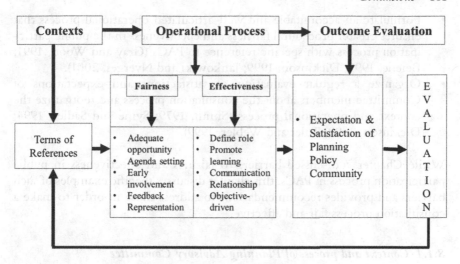

Figure 8.1 A conceptual model for PAC.

and wishes to maintain Garden City image, any DA needs to address ways of maintaining its image with greater input from the residents, community organizations and the development proponents. The Territory Plan and the National Capital Plan have also advocated retention of the 'Bush Capital' and 'Garden City' image.

The book proposes the conceptual model to achieve greater satisfaction of the wider public and Committee members, as they are the recipients of planning decisions. The outcome of this research does not expect the conceptual model to emerge as the perfect remedy, but recommends a process that will help achieve maximum performance regarding inclusion of fairness and effectiveness in public participation process of the PAC. Based on available literature (Beierle, 1999; Dickinson, 1999; Gray and Wood, 1991; Smith, 1987, 1993; Innes and Booher, 1999a, 1999b, 2000; Jankowski and Nyerges, 2001; Lauber and Knuth, 1998, 1999, 2000; Webler, 1995; Webler and Tuler, 2000) the proposed conceptual model presents two interrelated aspects to produce a process that is fair and effective, and develops the guidelines for evaluating existing processes. The guidelines describe the criteria of fairness and effectiveness, and related question sets for evaluating PACs. Some case-specific recommendations for improving consultation processes of PAC are also proposed. Based on the literature and the findings of this study, the following discussions summarize recommendations that will help achieve fairness and effectiveness in public participation process. The gist of the recommendations is as below.

- Develop a rational context with clear definition of pre-conditions for the PAC and its role into the planning decision (Gray and Wood, 1991; Smith, 1993; Dickinson, 1999).

- Formulate an accountable and well–articulated operational process that ensures greater procedural fairness and effectiveness in the public participation process with specific reference to PACs (Gray and Wood, 1991; Beierle, 1999; Dickinson, 1999; Jankowski and Nyerges, 2001).
- Organize a regular evaluation of satisfactions and expectations of Committee members about the consultation process and reorganize the context and operational process (Smith, 1979; Syme and Sadler, 1994; Dickinson, 1999; Tuler and Webler, 1999).

While Chapter 7 discussed barriers to fairness and effectiveness in public participation process in PACs, this chapter discusses specific examples of such barriers and provides recommendations to address those in order to make a consultation process fair and effective.

8.1.1 Context and process of Planning Advisory Committee

To design a fair and effective context of public participation process that is an integral part of a larger planning activity, a planning agency should initially address three major program elements: the terms of reference and its related activities, the public who might and should be on the Committee, and the issues the Committee would be dealing with. However, there are many other factors that need to be defined clearly into the terms of reference and its related activities: How the Committee is organized? How to identify major planning community to be involved? and How to formulate operational procedures of conducting committee meetings?

The terms of reference

The terms of reference for the Committee must be tailored to meet the needs of neighborhoods and other planning stakeholders, consistent with plans and guidelines. The role of PACs must also be addressed clearly in the terms of reference with statements of the Committee's role and purpose. The role should not be restricted as to prevent the Committee from contributing effectively in planning decisions. A Committee should be purpose-driven and everyone on the Committee must understand the purposes for which Committee is being set up and how it will achieve them. They should not feel ineffective in influencing the planning decisions.

However, some planners believe that the Committee's role should remain advisory and not be diverse or involve in every stage of the planning process. The excessive load for the Committees with many development applications may hinder them in achieving their main purpose of directions for their own neighborhoods. Committees should be formed only to comment at the strategic level of planning and development matters, instead of examining every development application regardless of its importance to the community. Thus, when formulating protocol, considerable, time and effort must be spent in defining the Committee's role, forms, functions and purposes.

Determining the balance between a strategic and issue-specific role of the Committee members is rather difficult until it is in the operational stage of the consultation process. One way to address this issue is to arrange regular reviews of the Committee's role by the planning agency to assess whether the Committee is effectively playing its role. In order to do that the planning agency may form a review panel comprising people from all the planning policy community. The function of the review panel should be to assess whether any amendments to the Committee's role are required to improve it. The amendments could therefore be incorporated within the existing terms of reference.

Membership of the Committee

Membership of the Committee is always a critical issue. When a PAC is decided on by the decision-makers as the type of public consultation body, one of the first steps is to determine the people who should be on the Committee. Webler (1992: 223) notes that 'if a balance is not achieved, or if overzealous supporters of interests are chosen then a large piece of the population may be shut out of the process'. Therefore, the balance of membership on the Committee is always in question. In the proposed model of PAC, all planning stakeholders – residents, developers, planning agencies, non-governmental organizations, as well as enthusiastic individuals with significant interests in planning and development matters – should have formal membership on the Committee.

To ensure a balanced membership on the Committee, members must acquire confidence in the wider public. A PAC consisting of planning and design experts or only community and resident representatives cannot achieve confidence of the planning agency and the wider community. A more balanced membership that includes both planning professionals and non-professional planning experts would better serve the needs of the planning policy community, and consequently be more likely to operate as an effective public participation mechanism. This recommendation could be contradictory to achieving greater democratic representation on the Committee; however, people in Canberra show little interests in regular participation, unless they are severely affected by the development programs. In such situation, representatives of cross-sections of people such as community leaders, church leaders and other representatives of community-based organizations should be included on the Committee to ensure greater representation and regular attendance at meetings.

This book recommends giving formal memberships to planners on the Committee as members, which may improve its effectiveness as a policy-input mechanism. The benefits are that it could improve communication by providing a direct link between the Advisory Committees and the Planning

Authorities. It would also improve credibility by giving the Advisory Committee a sense of authority and power. However, interview data indicate that residents and community groups do not support this recommendation; rather, they want a PAC with residents and community groups; however, they acknowledge the importance of planners attending the meetings regularly. Interviews with planners suggest that Committees should consist of people from diverse backgrounds such as youth, single mothers, the disabled, parents' and citizens'Committees, church leaders and community organizations.

Concern was also noticed about Ministerial appointees on PACs. Most residents and community groups do not support excessive appointees by the Minister for Planning. Interviews with the Committee members indicate that appointees are mostly developers, real estate personnel and representatives from big business organizations. The Minister never nominates a member from the community or from the residents' groups. Thus, a balance of Ministerial appointees would probably ease distrust among other members. Even though most of them oppose Ministerial appointees because of their backgrounds, appointments by the Minister or planning agency from a cross section of the residents and community organizations would make it balanced in its representation.

Balance and inclusiveness

Advisory Committees consist of people with diverse backgrounds including the residents, community, businessmen, technical experts and planning professionals. Identifying important stakeholders and ensuring their regular presence in the meetings is very important. A universal answer about which stakeholders to involve in management is neither possible nor desirable because the choice of stakeholders should depend both on the purpose of involvement and the contextual factors (Chase et al., 2000: 212). On the other hand, the members from resident and community groups on the Committee may raise the justifications of other representatives such as developer, businessman and architect on the Committee. This has been reflected in membership of the Committee, where the residents and community groups have strongly indicated that the number of business representatives and Ministerial appointees should be limited on every Committee. Appointees have created distrust among other members on evaluating development applications for the approval. Most appointees support development applications that propose more multi-unit development close to the central business district, but residents and community groups take on the opposite view. The conflicting interests of representatives often generate distrust among the Committee members. The business representatives and development proponents indicate that the number of community activists and environmental lobbyists should be limited, since they believe that those groups have an anti-development attitude which hinders the appropriate development for urban renewal projects. Apart from recognizing such conflicting attitudes to representation on the Committee, it is also important to ensure the involvement of underrepresented groups.

To ensure greater balance and inclusiveness, some guidelines are suggested for selection of the PAC: involve stakeholders with a broad range of attitudes, involve stakeholders with difference types of input to offer, involve traditionally underrepresented groups, and involve groups that have polarized views. In addition, special consideration should be given to the following selection issues:

- Select business representatives for a Committee who have established businesses in the area of concern, not those who only have interests in business in the area.
- Select development proponents' representatives who have enough understanding of the area of concern, not those who come only to serve the interest of the proponents; and
- Select appointees who understand the area of concern and are acceptable to other members, not those who neither live nor have businesses in the area.

Relationships among planning policy community

An effective consultation process ideally depends on good relationships among members of the Committee; distrust and hostility always hinder a consensus decision. As discussed earlier there were two groups on the Committee: the resident and community groups on the one hand, and the business representatives and Ministerial appointees on the other. The relationship between these two groups tends to be hostile and adversarial. Long-time members view that the consultation process is just to maintain planning obligations to consult planning policy community and has never been used as a mechanism to gather community input for planning decisions. Disillusion leads to a relationship which some members term adversarial and confrontational. As a result, both planning staff and the Committee members are unwilling to co-operate for better planning outcomes.

The relationship between planning staff and the Committee would benefit from more open and regular communication. This book suggests the ways to improve relationships in the PAC. If the Committee and planning staff review the Committee's objectives and issues of importance to the planners, both parties will be better informed of and sensitive to the concerns of others. To achieve better relationships in the policy community, the protocol should include provisions for regular consultation with planning staff outside the meetings. The meeting requirement would result in improved communication between Committee and planning staff and, enhance their relationships, which would in turn contribute to the effective operation of the consultation process.

The relationship between residents and development proponents is also important. Some Committee members blamed the proponents for not providing timely planning documents and language accessible information. A planning member on the Committee acknowledged that development proponents should employ people to prepare documents that laypeople on the Committee can understand. This does not happen in Committee; instead some planners and developers are critical of the Committee for inadequate understanding of planning and development matters.

Agenda setting and minute taking

The agenda is essential for the decision-making process. Decisions are reached through evaluating and discussing all agenda items: development applications and other issues put before the Committee members for their comments. The agenda should indicate the amount of time to be spent on each item. Webler and Tuler (2000) observe that an agenda can restrict participation by not allocating enough time for participants to speak and raise their concerns, and that sometimes it can also impair the quality of discussion on an issue by scheduling time in such a way that participants cannot attend the meeting in time. In PACs, there are many reasons to give importance to agenda setting, because identifying the meeting agenda is almost the same as defining the planning problems. In traditional models of public participation such as public meetings, workshops, information nights and surveys, people have the chance to put their concerns into the meeting agenda or raise them during the discussion. Unfortunately, in PACs, the meeting agenda is normally set by the planning agency either independently or in consultation with the Committee chair. In many cases, however, planning agency sets the agenda with the development proponents for the larger consultation processes, in which the Advisory Committee may be included. This practice normally leads to mistrust among others on the Committee. In Committee's consultations, for instance, members believe that the planning agenda is normally set by the development proponents in consultation with the planning staff, and the community has very little influence.

Some members of the Committees complaint that the planning agency sets the agenda based on available development applications and present these applications to the meetings, so the Committee has nothing to contribute to the agenda. Special attention and consideration should be given to ensure that all planning issues are included for discussion, and no planning issues should be officially excluded from the agenda. In addition, all members should have the opportunity to voice their concerns at every meeting and, enough time should be given to them. Members should not be given time at the end of the meeting while discussing other businesses, should declare if they have anything to add before the start of the formal meeting, and ample time should be allocated for discussion.

The time allocated for discussion of every item is very important. The Committee does not have any guidelines to allocate time for the discussion of the meeting agenda. The Committee Coordinator often makes an independent decision for allocating time and rarely consults with the convenor, but the coordinator may not know the importance of specific items and time requirement for discussion. There should be clear guidelines to allocate time for every agenda item in consultation with the Committee chair and other experienced members.

Sometimes discussion on some items goes beyond the time limits. Once time has been allocated rationally, it should be maintained rigidly. This process will fairly distribute the time for discussion and will also provide an equal chance for all participants. Special encouragement should be given to the wider public to speak and to raise their concerns in planning meetings. Public speaking rights should be ensured significantly at any open planning meetings.

The minutes of the PACs have been subject to little detailed research. There has been discussion of the importance of minute taking, which translates into the Committee's advice to the planning agency. The minutes should ensure that all concerns of the Committee have been recorded accurately and succinctly and should clearly state all the differences of opinion and expectations of the planning policy community discussed in the meetings. A Committee member indicates that development proponents may give more importance to 'traffic, access to civic amenities, parking, affordable housing, and urban renewal, while residents and community groups may give more attention to noise, smells, congestion, and the bad impact of urban consolidations'. The differing expectations of planning outcomes in the policy community will help the planning agency to structure the Committee's preferences and future planning directions. Similarly, the views of the wider public should be structured in a way that reflects their opinions and firm expectations through the consultation process.

The meeting minutes normally record individual comments on development applications. The members also show their disagreements on every issue and give opinions individually. On many occasions, consensus is reached, but does not emerge as a collective decision that can be put forward to the planning agency as community input. In such situation, the Committee should reach to a consensus with a majority of Committee members that can make firm decisions on planning and development matters.

Early involvement

A major barrier to the participation process is the lack of easy and early access to the pertinent information. This issue is particularly important in relation to information and notification of planning proposals and development applications. Interviews with Committee members show that early notification of development applications is essential. The notification should appear in a community-accessible form using simple language whenever possible and should have clear recognizable spatial information to avoid misunderstanding of explanations. The information should be made available to the affected members of the Committee. Large images of development sites should be displayed with clear information about the proposed development before formal applications are lodged. The meeting agenda and other important information should be published in local newspapers and posted on community notice boards, shopping complex and community centers. It is important to make use of local media to publicize information of public interest, and the planning agency and Committee should use the Internet, community bulletin boards and other forms of digital access. The planning agency should initiate processes to make information more accessible to the planning stakeholders including the wider public.

Adequate opportunity

This evaluative criterion encompasses many sub-criteria for members' opportunity for participation in the consultation process. They are:

- Equal access to information;
- Equal capacity to participate;
- Equal power to influence planning decisions;
- Equal opportunity to challenge the justification of planning proposals;
- Equal opportunity to challenge other member's claims;
- Access to information and knowledge;
- Access to planning documents;
- Adequate time allotment for reviewing development applications;
- Sufficient time for submission;
- Ample time for discussion; and
- Allocation of time for the wider public to discuss different issues at the meetings.

All Committee members interviewed commented on every criterion mentioned above but did not all agree with them. Business representatives and Ministerial appointees often said that other members on the Committee were given adequate opportunity in every aspect of the consultation process but they wanted more, which was not possible through consultation process. However, other members on the Committee indicated that they were not given adequate opportunity. The most pressing concerns were the lack of available time and language-accessible planning documents. This indicates that the Committees should be given enough time to review development applications for an informed discussion in the meetings. As some members may not have adequate understanding of planning and development matters, they should be provided with planning documents in accessible language, without using planning jargons that might create confusion in understanding. This initiative will give them an opportunity to participate in meaningful discussion and to debate the rationality of proposed development applications. A Committee member suggested that the planning agency and development proponents should employ social planners who can understand the community's social and cultural values and can use planning terms in such a way that laypeople on the Committee may understand as well, and contribute more to the process of development approval. However, planners with an architecture background, along with other physical planners, dominate planning agencies in most cities in Australia (Kaufman, 1999). It is noteworthy that there are no social planners in senior positions in the ACT's Planning Authority. Planning education does little to introduce social values to the architects and planners. Gleeson and Low (2000) observe that Australian planning education seems not to be interested in introducing more coherent social, economic and environmental issues to be included in the planning curriculum. However, it seems that social and environmental planners must inevitably take part in producing planning documents for the community and other planning stakeholders.

Communication, feedback and responsiveness

Communication and feedback between Committee members and planning staff must be efficient and responsive. Once the Committee receives a full and logical account of reasons as to why its recommendations and advice were accepted, modified or rejected, it is better able to understand planning policy and priorities. Lack of such communication has caused much resentment. An improved situation is inevitable to make feedback and responsiveness more effective and consistent, so that there will be a sense of accountability in the Committee, who will feel that they are being heard and able to influence the planning decisions.

Promote learning and information session program

Members of the expert Advisory Committees are more-or-less well informed on relevant issues and have high professional expertise in their own fields (Ashford, 1984). Thus, they are designed to provide technical and high-level strategic advice on issues relevant to the agency's goals, such as nature of strategic plans and identification of appropriate timelines for reviewing planning laws and design guidelines. Non-professional members and lay people are also included on many Advisory Committees as affected citizens to capture their input. To keep all these Committee members up to date with recent changes in laws, government executive orders and legislative amendments, a regular or session-specific outreach program would be beneficial. PAC members would be informed of current changes in planning laws, design guidelines, land use zoning, and particularly political agendas that might affect planning and design matters.

An outreach program is also necessary to make participants competent enough to argue and establish claims to validity and assertions made on relevant matters (Ashford and Rest, 1999; Webler, 1995). Members must be well informed on planning laws and guidelines to articulate their concerns effectively and accurately.

Planning staff also require adequate communication skills for effectively conducting a consultation. Planning staff have the necessary planning education, but often lack skill in communicating with the planning policy community. A Committee member observed that planners at Planning Authority were good at explaining physical planning but failed to address social values to put into planning practices. Kaufman (1999) advocates introducing sociological theory into the Australian planning education to enhance the planners' communication skills and understanding of the social values of the community living within a neighborhood.

The lack of available neighborhood profiles is another issue with Committee members. A basic introduction of a neighborhood profile to the Committee members in the outreach program will ensure informed and timely comment. The Committee members are not given prior information on the neighborhoods, so they feel discouraged to participate. As mentioned, Planning Authority has prepared six neighborhood snapshots, which are a hopeful sign that neighborhood indicators will be provided to unveil community needs and future directions. Long and Beierle (1999) commented that the advisory groups should have two educational objectives: educating participants on the Committee and educating the wider public outside the Committee, coupled with discussion during the meeting. Most of the Committee members agreed that an educational program would be beneficial; however, a Committee is a relatively small organization and has no funding to receive community input. There should be a provision for education to ensure wider community participation in planning decisions. Long and Beierle (1999) point out that the Advisory Committee is a source not only of values, assumptions and preferences, but also of facts and to generate alternatives.

8.1.2 Organize outcome evaluation of the participation process

One of the objectives of this book is to develop criteria for evaluating fairness and effectiveness of public participation process in PAC. The proposed criteria should satisfy most of the planning policy community and should be a fair and legitimate way to incorporate community and individual values into planning decisions. After rigorous evaluation of the consultation process of PACs and planning outcomes, this book provides two meta-criteria, ten evaluative criteria, twenty-six sub-criteria and sixty-four question sets (Tables 8.1 and 8.2) to evaluate public participation, particularly its procedural fairness and effectiveness in PAC. The proposed criteria and question sets reflect Webler's (1992, 1995) theory of fairness and competence, which was applied in waste management programs (Webler, 1992, 1995), forest management programs (Webler and Tuler, 2000) and nuclear waste clean-up programs (Webler, 1999), but this book developed criteria that are appropriate for evaluating PAC.

Table 8.1 Evaluative criteria for PACs: Fairness

A. Guidelines for evaluating fairness criteria: adequate opportunity
A1. The PAC should have equal opportunity to participate meaningfully.

- Does the Planning Authority have the relevant planning information readily available at the committee's disposal (e.g. updated planning laws, rules, state of environment reports, design guidelines, spatial data and updated *Land Acts* that affect planning and development matters)?
- Is the PAC provided with language-accessible documents?

A2. The PAC should have equal opportunity for access to information.

- Is the PAC being provided with detailed neighborhood indicators before meetings?
- Does the Planning Authority or developers provide relevant information to the PAC before the meeting?
- Is the committee satisfied with the methods and techniques used for information dissemination?

A3. The PAC should have equal power to influence planning decision.

- Do the PAC members feel that their recommendations have been taken into consideration?
- Do the PAC members feel that they are equal partners in planning decision?
- Does the PAC discourage excessive use of the authority's discretionary power to override PAC recommendations in taking the final decision?
- Is the PAC given any legal privileges to lodge complaints with an administrative appeal tribunal, if their recommendations are not given consideration in the planning decisions?

B. Guidelines for evaluating fairness criteria: representation
B1. The PAC should ensure greater representation from all the planning policy community.

- Does the PAC equally represent by the existing planning policy community?
- Is the PAC free from excessive political leaders on the committee?
- Is the PAC free from unknown representatives, who only maintain the routine work of their organizations?
- Are the PAC members normally elected or appointed believed to have suitable personalities, moral standards and planning understanding in the areas?

C. Guidelines for evaluating fairness criteria: feedback and responsiveness
C1. The PAC should perceive that they have received adequate feedback from the Planning Authority.

- Does the PAC receive regular feedback from the Planning Authority?

C2. The PAC should receive explanations for acceptance, rejection and modification of recommendations.

- Does the PAC receive justification and adequate explanation why their comments were accepted, modified or rejected?
- Does the consultation process have provisions to reply to and challenge the assertions of authorities relating to the rejection of PAC advice?
- Does the process allow the PAC to be one of the affected parties to seek legal solutions on disagreements?

(continued)

Table 8.1 (Continued)

D. Guidelines for evaluating fairness criteria: agenda setting and minute taking

D1. The process should have provision for all PAC members to define planning and development issues before formal development applications are lodged.

- Does the process give the PAC the opportunity to define planning issues for the decision-making?
- Does the process allow PAC to question planners and designers and shape the agenda for planning decisions?

D2. The process should have provision to include all matters for the PAC agenda.

- Does the process include all planning and development matters in the PAC agenda?

D3. The meeting minutes should include comments from all planning stakeholders

- Does the process provide the PAC, the wider public and development proponents with an equal opportunity to include their concerns into the meeting minutes?

D4. The meeting minutes should available for public perusal and should be distributed among planning stakeholders in good time.

- Are the minutes publicly available for discussion and access to the wider community?
- Are the minutes distributed to all members in good time?

E. Guidelines for evaluating fairness criteria: early involvement

E1. The PAC should have the opportunity to be involved in inter-departmental planning meetings.

- Does the PAC have adequate opportunity to be involved early in the inter-departmental meetings before formal development applications are initiated?

E2. The process should make opportunities for the PAC members to be involved during the conceptualization of the proposal.

- Does the PAC have the opportunity to be involved during the conceptualization of development proposals?
- Does the PAC have the opportunity to put forward its ideas and concerns before the conceptualization the proposed development applications?
- Does the PAC have the opportunity to check whether the development applications comply with planning and design guidelines and committee recommendations?
- Does the process provide an opportunity to everyone on the PAC to define and to determine the advantages and disadvantages of proposed development applications?
- Does the process allow the PAC to have access to other stakeholder definitions and conceptualization of development applications?

Source: Adapted from Webler, (1992, 1995, 1999); Beierle (1998); Octeau (1999); Chess (2000); Rowe and Frewer (2000) Webler and Tuler (2000)

8.3 Applications of evaluative criteria

The proposed evaluative criteria and question sets can be applied to a wide array of citizens' advisory groups but are probably most appropriate for planning and environmental advisory groups. Manifestations of the proposed model in specific contexts may require adjustments in context or order. Unlike statistical analysis, which can be proved or disproved by hypothesis

Table 8.2 Evaluative criteria for PACs: Effectiveness

A. Guidelines for evaluating effective criteria: defined role
A1. The terms of reference should clearly define the role of all the planning policy community.

- Do the terms of reference clearly define the tasks, roles and functions of every discourse participants?
- Does the planning authority clearly define the objectives of consultation with the PAC in the terms of reference?

A2. All PAC members should have clear understanding about the terms of reference and outcomes of the consultation process in the planning decision.

- Do the PAC members understand their defined roles and are they determined to carry them out effectively?

A3. The PAC members should be competent to play their defined role in the planning decisions.

- Do the PAC members have adequate understanding of planning terms, definitions and design matters to advise the planning authority?
- Is the PAC able to articulate its concerns competently and effectively?

B. Guidelines for evaluating effective criteria: promote learning
B1. The process should have effective outreach programs for the members about the planning laws and design guidelines that are essential to comply with development applications.

- Does the planning authority organize outreach programs to introduce planning laws, design guidelines and the process of development approval to the PAC members?
- Does the planning authority organize outreach programs for the incumbent planners so they can carry out the consultation process effectively and succinctly?
- If the outreach program is organized, does it give adequate information about neighborhood indicators to the PAC members?
- Does the process keep the PAC members up to date about current changes in planning laws and design guidelines at present and possible changes in near the future?

B2. The process should have a clear plan for public participation that includes a way for learning to the PAC members.

- Does the planning authority evaluate its existing context and process to promote a way of further learning to improve its context and process?

C. Guidelines for evaluating effective criteria: communication
C1. The PAC should have regular communication with all the planning policy community.

- Do the PAC members have effective communication internally with other representatives?
- Does the process provide PAC members with an opportunity to informally discuss their feelings with other members before formal discussion in the meetings?
- Does the PAC have regular communication with the wider public, voluntary organizations and political parties to discuss planning policy?
- Does the PAC chair communicate with other members as well as planning stakeholders effectively and punctually?

(continued)

Table 8.2 (Continued)

C2. The PAC should have a formal structure to communicate with others in the planning policy community.

- Does the process have provision for regular communication with the planning and support staff?
- Does the PAC maintain a formal structure to communicate with planning and support staff at the planning authority?
- Does the PAC maintain a common listserv for electronic communication with other planning stakeholders?
- Do the planning authority and support staff respond punctually to the PAC's communications on planning and development matters?

C3. The PAC should use appropriate mechanisms to effectively communicate through spatial planning data.

- Do the development proponents use 3D models for displaying development applications during the PAC meetings?
- Does the planning authority provide development applications and neighborhood information to the Web for PAC members?
- Is the PAC facilitated by the Web-based GIS access to make comments on development applications?

C4. The process should provide adequate financial and human resources to the PAC for effective communication.

- Does the PAC receive adequate funding to collect community input, conduct its own research and manage its office activities?
- Does the planning authority provide adequate secretarial support?

D. Guidelines for evaluating effectiveness criteria: relationship and trust.
D1. The PAC members should have amicable relations with each other on the committee.

- Does the PAC maintain an amicable relationship among all members on the committee?
- Does the chair maintain a good relationship among all members on the committee?
- Does the PAC maintain a responsive relationship with planning and support staff in the planning agency?

D2. The PAC should maintain amicable relationships with others in the policy community.

- Is the PAC believed to have an amicable relationship with the development proponents?
- Does the meeting process establish rules about acceptable behavior at meetings, to be strictly enforced?
- Does the meeting process establish rules to avoid personal attacks during the discussion?

D3. The PAC member should have trust in the chair conducting the meeting.

- Do the PAC and development proponents trust the committee chair?
- Does the process indicate the existence of trust among all members regardless of whether they are elected or appointed?

Table 8.2 (Continued)

E. Guidelines for evaluating effectiveness criteria: objective driven

E1. The process should have provision to evaluate its authority's objective and the satisfaction of the PAC members.

- Does the process have any time to evaluate its success or failure and future directions of the committee?
- Does the PAC show satisfaction with the planning outcomes through the consultation process?
- Have the expectations of PAC members been adequately met through the consultation process?

E2. The PAC should have a provision to monitor its advice being considered in planning decisions.

- Does the PAC monitor whether its advice is considered by the planning authority?
- Does the PAC form sub-committees, when necessary, to monitor its advice being adequately addressed by the planning authority?

Source: Adapted from Webler, (1992, 1995, 1999); Beierle (1998); Octeau (1999); Chess (2000); Rowe and Frewer (2000) Webler and Tuler (2000)

testing, the proposed evaluation criteria and question sets must be treated with a degree of interpretation and flexibility, since they may not be appropriate to other types of planning and environmental advisory groups. However, all are believed to be consistent with the objective of achieving a fair and effective public participation process in PACs.

8.4 Recommendations for further research

The task of studying public participation process is complex, and evaluation of the existing process is much more difficult as it requires individual interpretations and reactions to others behavior on the issues commonly shared by the participants. This book has evaluated the fairness and effectiveness criteria of the consultation process and developed a conceptual model for PACs. The conceptual model suggests some evaluative criteria and question sets. The model can be applied in Australia and elsewhere, particularly in planning and environmental advisory groups. The book suggests the following directions for the future research.

1 There is a need for further research on the Committees' input into planning decisions. This will allow the researchers determine whether PAC input is accepted and used in the final planning decisions. Further study should attempt to clarify whether PAC input is rejected because all PAC input is routinely ignored, or if PAC input is rejected based on justifiable planning reasoning.

2 Further research could determine how planning authority structures the Committee's advice. This would allow the researchers understand the

competency of the planning authority to analyse thoroughly the committee input for the final decisions.

3 Further research also needs to be carried out to measure the gap in expectation between the community and the planning agency on decision outcomes. A survey could therefore be conducted to determine the rationale of differing expectations among the planning policy community and suggest ways of bridging the gap. This would formulate a strategic approach to understanding the expectations of planning stakeholders and to meeting those expectations in a way that all parties feel empowered.

4 A study assessing the feedback and responsiveness between the PAC and planning agency is also recommended. This will allow the researchers to evaluate the minute taking process and to understand how PAC members contact the planning department and receive no feedback in time. All meeting minutes indicate that the members contact many ACT departments for relevant information and make pertinent queries, but departments normally do not reply to the members, unless asked repeatedly.

References

Abbott, J. (1996). *Sharing the city: Community participation in urban management.* Earthscan Publications Limited. London.

ACT Legislative Assembly (2000). *Hansard of the ACT Legislative Assembly: Committee for Planning and the Environment.* [Online], 29 July 2002, http://www.hansard.act.gov.au/start.htm.

ACT Planning and Land Authority (ACTPLA) (2003a). *ACT planning and land authority: Its role in the planning & land management systems.* ACT Government. Canberra.

ACT Planning and Land Authority (ACTPLA) (2003b). *Planning and land reforms in the ACT: Main changes and how they work.* ACT Government. Canberra.

ACT Planning and Land Authority (ACTPLA) (2003c). *Development application assessment: Changes to decision-making and review processes.* ACT Government. Canberra.

ACT Planning and Land Authority (ACTPLA) (2003d). *Planning and land council: Its role in the planning & land management systems.* ACT Government. Canberra.

ACT Planning and Land Authority (ACTPLA) (2003e). *Land development agency: Its role in the planning & land management systems.* ACT Government. Canberra.

Alterman, R., Harris, D., and Hill, M. (1984). The impact of public participation on planning. *Town Planning Review,* 55(2), 177–196.

Altheide, D., and Johnson, J. M. C. (1998). Criteria for assessing interpretive validity in qualitative research. In N. K. Denzin and Y. S. Lincoln (eds.), *Collecting and interpreting qualitative materials,* Sage Publications. Thousand Oaks, CA.

Arnstein, S. (1969). A ladder of citizen participation. *Journal of the American Institute of Planners,* 35(4), 216–224.

Ashford, N. A. (1984). Advisory committees in OSHA and EPA: Their use in regulatory decision-making. *Science, Technology, and Human Values,* 9(1), 72–82.

Ashford, N. A., and Rest, K. M. (1999). *Public participation in contaminated communities. center for technology.* Policy and Industrial Development, MIT. Cambridge, MA.

Australian Bureau of Statistics (ABS) (1999). *Use of the Internet by householders in Australia.* Cat. No. 226. Canberra.

Australian Labor Party (2001). *Labor's plan to maintain the garden city.* Fact Sheet No. 4. Canberra.

Banfield, E. C. (1961). *Political influence: A new theory of urban politics.* Free Press. New York.

Barndt, M. (1998). Public participation GIS–barriers to implementation. *Cartography and Geographic Information Systems,* 25(2), 105–112.

Barndt, M., and Craig, W. (1994). Data providers empower community GIS efforts. *GIS World,* 7(7), 49–51.

Bazeley, P., and Richards, L. (2000). *The NVivo qualitative project book*. Sage Publications. London.

Beder, S. (1999). Public participation or public relations? In B. Martin (ed.), *Technology and public participation*, Science and Technology Studies, University of Wollongong, Wollongong, 143–158.

Beierle, T. C. (1998). *Public participation in environmental decisions: An evaluative framework using social goals*. Resource for the future. Washington, D.C.

Beierle, T. C. (1999). Using social goals to evaluate participation in environmental decisions. *Policy Studies Review*, 16(3/4), 75–103.

Berg, B. L. (1989). *Qualitative research methods in the social science*. Allyn & Bacon. MA.

Birkeland, J. (1999). Community participation in urban project assessment: An ecofeminist analysis. In B. Martin (ed.), *Technology and public participation*, Science and Technology Studies, University of Wollongong, Wollongong, 113–142.

Brager, G., Specht, H., and Torczyner, L. (1987). *Community organization*. Columbia University Press. New York.

Brody, S. D., Godschalk, D. R., and Burby, R. J. (2003). Mandating participation in plan making: Six strategic planning choices. *Journal of the American Planning Association*, 69(3), 245–264.

Bryman, A. (1988). *Quantity and quality in social research*. Unwin Hyman. London.

Buchy, M., and Hoverman, S. (1999). *Understanding public participation in forest planning in Australia*. ANU Forestry Occasional Paper, 99.2. The Australian National University. Canberra.

Burawoy, M. (1991). The extended case method. In M. Burawoy (ed.), *Ethnography unbound: Power and resistance in the modern metropolis*, University of California Press. Berkely.

Burke, E. M. (1979). *A participatory approach to urban planning*. Human Science Press. New York.

Byrne, J., and Davis, G. (1998). *Participation and the NSW policy process: A discussion paper for the cabinet office new South Wales*. New South Wales Cabinet Office. Sydney.

Caddy, J., and Vergez, C., (2001). *Citizens as partners: Information, consultation and public participation in policy making*. Report prepared for Organization for Economic Cooperation and Development. Paris.

Cameron, J. (2000). Focusing on the focus group. In I. Hay (ed.), *Qualitative reserach methods in human geography*, Oxford University Press. Melbourne.

Campbell, H., and Marshall, R. (2000). Public involvement and planning: Looking beyond the one to the many. *International Planning Studies*, 5(3), 321–344.

Canadian Council of Forest Ministers (1997). *Criteria and indicators of sustainable forest management*: Technical Report 1997, Cat. Fo75-3/6-1997E. Ottawa.

Carson, L. (1996). How do decision makers in local government respond to public participation? PhD thesis, Faculty of Education, Work and Training, Southern Cross University. Lismore.

Carson, L. (1999) Random selection: Achieving representation in planning. *Alison Burton memorial lecture*, Royal Australian Planning Institute. Canberra, 31 August 1999.

Catanese, A. J. (1984). *The politics of planning and development*. Sage Publications. Beverly Hills, CA.

Chambers, R. (1997). *Whose reality counts? Putting the last first*. IT Publications. London.

Chase, L. C., Schusler, T. M., and Decker, D. J. (2000). Innovations in stakeholders involvement: What's the next step. *Wildlife Society Bulletin*, 28(1), 208–217.

Checkoway, B. (1981). The politics of public hearings. *The Journal of Applied Behavioral Sciences*, 17(4), 566–582.

Chess, C. (2000). Evaluating environmental public participation: Methodological questions. *Journal of Environmental Planning and Management*, 43(6), 769–784.

Cole, J. M., and Cole, M. F. (1983). *Advisory councils: A theoretical and practical guide for program planners*. Prentice Hall. Englewood Cliffs, New Jersey.

Cole, R. L. (1974). *Citizen participation and the urban policy process*. Lexington Books. Toronto.

Cole, R. L., and Caputo, D. A. (1984). The public hearing as an effective citizen participation mechanism: A case study of the general revenue sharing program. *American Political Science Review*, 78: 404–416.

Coleman, W. D. (1985). Analysing the associative action of business: Policy advocacy and policy participation. *Canadian Public Administration*, 28(3), 413–433.

Conrad, E., Cassar, L., Christie, M., and Fazey, I. (2011). Hearing but not listening. A participatory assessment of public participation in planning. *Environment and Planning C: Government and Policy*. 29: 761–782.

Corbin, J., and Strauss, A. (1990). Grounded theory research: Procedures, canons and evaluative criteria. *Qualitative Sociology*, 13(1), 3–21.

Cormick, G., Dale, N., Edmond, P., and Stuart, D. (1996). *Building consensus for a sustainable future: Putting principles into practices*. National Round Table on the Environment and the Economy. Ottawa.

Craig, W., Harris, T., and Weiner, D. (eds.) (1999). *Empowerment, marginalization and public participation GIS*. NCGIA. Santa Barbara, CA.

Craig, W. J., and Elwood, S. A. (1998). How and why community groups use maps and geographic information. *Cartography and Geographic Information Systems*, 25(2), 95–104.

Creighton, J. (1986). *Managing conflicts in public involvement settings: Training manual for Bonneville power administration*. US Department of Energy. Washington, D.C.

Creighton, J. (1993). *Guidelines for establishing citizen's advisory groups*. US Department of Energy. Washington, D.C.

Creswell, J. (1994). *Research design: Qualitative and quantitative approaches*. Sage Publications. Thousand Oaks, CA.

Creswell, J. (1998). *Qualitative inquiry and research design: Choosing among five traditions*. Sage Publications. Thousand Oaks, CA.

Creswell, J. and Creswell, D., (2017). *Research design: Qualitative, quantitative and mixed methods approaches*. Sage Publications. Thousand Oaks, CA.

Crosby, N. (1991). *Citizen's juries as a basic democratic reform*. Minneapolis. Jefferson Center.

Crosby, N., Kelly, J., and Schaefer, P. (1986). Citizen panels: A new approach to citizen participation. *Public Management Forum*, 46: 170–179.

Dandekar, H. (ed.) (1982). *The planner's use of information*. Hutchinson Ross Publishing Company. Pennsylvania.

Davies, J. C. (1998). *Public participation in environmental decision-making and the federal advisory committee act*. Centre for Risk Management, Resource for the Future. Washington D.C.

Davis, G. (1996). *Consultation, public participation and the integration of multiple interests into policy making*. OECD. Paris.

Day, D. (1997). Citizen participation in planning process: An essentially contested concept. *Journal of Planning Literature*, 11(3), 421–434.

Department of Urban Services (DUS) (2001). *Local area planning advisory committees: Guide for LAPAC members*. ACT Government. Canberra.

Department of Urban Services (DUS) (2002). *Annual report 2001*. ACT Government. Canberra.

DeSario, J., and Langton, S. (1987a). Citizen participation and technology. In S. DeSario and S. Langton (eds.), *Citizen participation in public decision-making*, Greenwood Press, Westport CT.

DeSario, J., and Langton, S. (1987b). Toward a metapolicy for social planning. In J. DeSario and S. Langton (eds.), *Citizen participation in public decision-making*, Greenwood Press. Westport CT.

Dickinson, W. H. (1999). Municipal environmental advisory groups: Form, function and effectiveness. PhD Thesis, Department of Geography, The University of Western Ontario. Ontario.

Donaghy, W. C. (1984). *The interviews: Skills and applications.* Foresman and Company. Glenview, ILL.

Driel, H. V. (2001). *Digitaal communiceren.* Uitgeverij, Boom te Amsterdam. Cited in Ghose, R., and Huxhold, W. (2001). The role of local contextual factors in building public participation GIS: The Milwaukee experience. *Cartography and Geographic Information Science*, 28(3), 195–208.

Dunn, K. (2000). Interviewing. In I. Hay (ed.), *Qualitative research methods in human geography*, Oxford University Press. Melbourne, 50–81.

Elwood, S., and Leitner, H. (1998). GIS and community-based planning: Exploring the diversity of neighborhood perspectives and needs. *Cartography and Geographic Information Systems*, 25(2), 77–78.

Elwood, S. A. (2000). Information for change: The social and political impacts of geographic information technologies. PhD Thesis, University of Minnesota. Minnesota.

Environment ACT (1999). *ACT environment advisory committee: Annual report.* ACT Government. Canberra.

Environment, Planning and Sustainable Development Directorate (EPSDD) (2020). *Pre-DA community consultation guidelines for prescribed developments: August 2020.* The ACT Government. Canberra.

Ertel, M. O. (1979a). A survey research evaluation of citizen participation strategies. *Water Resources Bulletin*, 15(4), 757–762.

Ertel, M. O. (1979b). The role of citizen advisory groups in water resources planning. *Water Resources Bulletin*, 15(6), 1515–1523.

Fagence, M. (1977). *Citizen participation in planning.* Pergamon, Oxford.

Faludi, A. (1973). *Planning theory.* Pergamon, Oxford.

Farrell, G., Melin, S., and Stacey, D. (1976). *Involvement: A Saskatchewan perspective.* Consulting Group Limited. Saskatchewan.

Field, P. A., and Morse, J. M. (1991). *Nursing research: The application of qualitative approaches.* Chapman & Hall. London.

Filyk, G., and Cote, R. (1992). Pressure from the inside: Advocacy groups and the environmental policy community. In R. Boardman (ed.), *Canadian Environmental policy: Ecosystems, polities and process*, Oxford University Press, Toronto.

Fiorino, D. J. (1990). Citizen participation and environmental risk: A survey of institutional mechanisms. *Science, Technology & Human Values*, 15(2), 226–243.

Fisher, K. F. (1984). *Canberra: Myths and models.* Institute of Asian Affairs. Hamburg.

Flick, U. (1998). *An introduction to qualitative research.* Sage Publications. London.

Foddy, W. H. (1993). *Constructing questions for interviews and questinnaires: Theory and practice in social research.* Cambridge University Press. Cambridge.

Fontana, A., and Frey, H. (1994). Interviewing: The art of science. In N. Denzin and Y. Lincoln (eds.), *Handbook of qualitative research*, Sage Publications. Thousands Oaks, CA.

Forester, J. (1999). *Planning in the face of power.* University of California Press. Berkeley.

Friedman, B. J., and Kaplan, M. (1975). *The politics of neglect: Urban aid from model cities to revenue sharing.* MIT Press. Cambridge, MA.

Game, A. (1988). Canberra And nation. In P. Grundy (ed.), *Canberra: A people's capital,* Australian Institute of Urban Studies. Canberra.

Gay, B. (1989). *Collaborating: Finding common ground for multiparty problems.* Jossey-Bass. San Franscisco.

Ghose, R., and Huxhold, W. (2001). The role of local contextual factors in building public participation GIS: The Milwaukee experience. *Cartography and Geographic Information Science,* 28(3), 195–208.

Ghose, R., and Huxhold, W. (2002). The Role of Multi-scalar GIS-based Indicator Studies in Formulating Neighborhood Planning Policy. *URISA Journal,* 14(2), 5–17.

Glaser, B., and Strauss, A. L. (1967). *The discovery of grounded theory: Strategies for qualitative research.* Aldine Publishing Company. Chicago.

Glass, J. J. (1979). Citizen participation in planning: The relationship between objectives and techniques. *Journal of the American Planning Association,* 45(2), 180–189.

Gleeson, B., and Low, N. (2000). *Australian urban planning: New challenges, new agendas.* Allen and Unwin. Sydney.

Goode, W. J., and Hatt, P. K. (1982). *Methods in social research.* McGraw-Hill. NY.

Grant, J. (1994). *The drama of democracy: Contention and dispute in community planning.* University of Toronto Press. Toronto.

Gray, B., and Wood, D. J. (1991). Collaborative alliances: Moving from practice to theory. *Journal of Applied Behavioral Science,* 27(1), 3–22.

Greenwood, T. (1984). *Knowledge and discretion in government regulations.* Environmental Protection Agency. Washington, D.C.

Greg, B., and Chin, S. Y. W. (2013). Assessing the effectiveness of public participation in neighbourhood planning. *Planning Practice & Research,* 28(5), 563–588.

Guba, E. G., and Lincoln, Y. S. (1994). Competing paradigms in qualitative research. In N. Denzin and Y. Lincoln (eds.), *Handbook of qualitative research,* Sage Publications. Thousand Oaks, CA.

Guglielmo, A. T. (1998). Nuclear waste, democracy, and risk: A procedural evaluation of the Columbia River Comprehensive Impact Assessment. Master's Thesis, University of Washington. Washington D.C.

Gunton, T. J. (1984). The role of the professional planner. *Canadian Public Administration,* Fall 1984, 399–417.

Habermas, J. (1979). *Communication and the evolution of society.* Beacon Press. Boston.

Habermas, J. (1984). *Theory of communicative action: Reason and the rationalization of society,* Vol. 1. Beacon Press. Boston.

Habermas, J. (1987). *The theory of communicative action,* Vol. 2. Beacon Press. Boston.

Hakim, C. (1987). *1. Research design: Strategies and choices in the design of social research.* Allen and Unwin, London.

Hall, J., and Stevens, P. (1991). Rigor in feminist research. *Advances in Nursing Science,* 13(3), 16–29.

Hammersley, M. (1992). *What's wrong with ethnography: Methodological explorations..* Routledge. London.

Hampton, W. (1977). Research into public participation in structure planning. In W.R.D. Well and J.T. Coppock (eds.), *Public participation in planning,* John Wiley & Sons. London.

Hannah, S., and Lewis, H. (1982). Internal citizen control of locally initiated citizen advisory committes: A case study. *Journal off Voluntary Action Research,* 11, 39–52.

Harding, R. (1998). *Environmental decision-making: The role of scientists, engineers and the public.* Federation Press Leichhardt, Sydney.

Harris, T., Weiner, D., Warner, T., and Levin, R. (1995). Pursuing social goals through participatory GIS: Readdressing South Africa's historical ecology. In J. Picknes (ed.), *Ground truth: The social implications of geographical information systems,* Guilford Press, NY, 196–222.

Heberlein, T. (1976). Some observations on alternative mechanisms for public involvement: The hearing, public opinion poll, the workshop and the quasi-experiment. *Natural Resources Journal,* 16: 195–212.

Heckman, L. A. (1999). *Methodology matters: Devising a research program for investigating PPGIS in collaborative neighborhood planning.* [Online], 5 May 2000. http://www.ncgia. ucsb.edu/varenius/ppgis/papers/heckman/heckman.html.

Homenuck, P. (1977) Evaluation of public participation programs. *Proceedings of the Canadian conference on public participation,* Alberta, Spring, 103–119.

Houghton, D. G. (1988). Citizen advisory boards: Autonomy and effectiveness. *American Review of Public Administration,* 18(3), 283–296.

Hutcheson, J. D. (1984). Citizen representation in neighborhood planning. *American Planners Association Journal,* 50(Spring), 183–193.

Hutcheson, J. D., and Prather, J. E. (1988). Community mobilization and participation in the zoning process. *Urban Affairs Quarterly,* 23(3), 346–348.

Innes, J. E., and Booher, D. E. (1999a). Consensus building as role playing and bricolage: Towards a theory of collaborative planning. *Journal of the American Planning Association,* 65(1), 9–26.

Innes, J. E., and Booher, D. E. (1999b). Planning institutions in the network society: Theory for collaborative planning. In S. Willet and A. Faludi (eds.), *Revival of strategic spatial planning,* Royal Netherlands Academy of Science. Amsterdam.

Innes, J. E., and Booher, D. E. (2000). Public participation in planning: New strategies for the 21st century. Paper presented at the annual conference of the Association of Collegiate Schools of Planning, 2–5 November 2000, CA.

James, R. F. (1999) Public participation in environmental decision-making: New approaches. *Paper presented at 12th annual national conference of the Environmental Institute of Australia.* Hobart, Tasmania.

James, R., and Blamey, R. (1999a). Application of the citizens' jury method to environmental management and environmental valuation. Paper presented at the workshop on public participation in environmental decision-making: A new Approach, using citizens' juries, 13 April 1999, The Australian National University. Canberra.

James, R. F., and Blamey, R. K. (1999b). Public participation in environmental decision making—rhetoric to reality. In International Symposium on Society and Resource Management, 7–10 July 1999. Brisbane.

James, R. F., and Blamey, R. K. (1999c). Citizen participation—some recent Australian developments. In International Conference on Pacific Science Congress, 4–9 July 1999. Sydney.

Jankowski, P., and Nyerges, T. (2001). *Geographic information systems for group decision-making: Towards a participatory geographic information science.* Taylor & Francis. London.

Kaufman, P. E. (1999). Conserving urban cultural landscapes: A critical examination of social values in landuse planning decisions. PhD thesis, Faculty of Environmental Design, University of Canberra. Canberra.

Kazar, A. (2003). More consultation call on development. *The Northside Chronicle,* 24(31), 3. 5 August, 2003, Canberra.

Kellogg, W. (1999). From the field: Observations on using GIS to develop a neighborhood environmental system for community-based organizations. *Journal of the Urban and Regional Information Systems Association*, 11(1), 15–32.

King, C. S., Feltey, K. M., and Susel, B. O. (1998). The question of participation: Toward authentic public participation in public administration. *Public Administrative Review*, 58(4), 317–326.

Kingston, R., Carver, S., Evans, A., and Turton, I. (2000). Web-based public participation geographical information systems: An aid to local environmental decision making. *Computers, Environment and Urban Systems*, 24(2), 109–125.

Knaap, G., Debra, M., and Olshansky, R. (1998). Citizen advisory groups in remedial action planning: Paper tiger or key to success. *Journal of Environmental Planning and Management*, 41(3), 337–354.

Krefting, L. (1990). *Rigor in qualitative research: The assessment of trustworthiness*. School of Rehabilitation Therapy, Queen's University. Ontario.

Krueger, R. A. (1994). *Focus groups: A practical guide for applied research*. Sage Publications. Thousand Oaks, CA.

Kvale, S. (1996). *Interviews: An introduction to qualitative research interviewing*. Sage Publications. New Delhi.

Kweit, M. G., and Kweit, R. W. (1981). *Implementing citizen participation in a bureaucratic society: A contingency approach*. Praeger. New York.

Laird, F. N. (1993). Participatory analysis, democracy, and technological decision-making. *Science, Technology & Human Values*, 18(3), 341–361.

Landre, B. K., and Knuth, B. A. (1993a). The role of agency goals and local context in Great Lakes water resources public involvement programs. *Environmental Management*, 17(2), 153–166.

Landre, B. K., and Knuth, B. A. (1993b). Success of citizen advisory committees in consensus based water resources planning in the great lakes basin. *Society and Natural Resources*, 6(3), 229–257.

Langton, S. (1981). Evolution of a federal citizen involvement policy. *Policy Studies Review*, 1: 369–378.

Lauber, T. B., and Knuth, B. A. (1998). Refining our vision of citizen participation: Lessons from a moose reintroduction proposal. *Society and Natural Resources*, 11: 411–424.

Lauber, T. B., and Knuth, B. A. (1999). Measuring fairness in citizen participation: A case study of moose management. *Society and Natural Resources*, 11: 19–37.

Lauber, T. B., and Knuth, B. A. (2000). *Citizen participation in natural resource management: A synthesis of HDRU research*. HDRU Series No. 00-7. Department of Natural Resources, Cornell University. Ithaca, NY.

Leining, M. (1994). Evaluating criteria and critique of qualitative research studies. In M. Morse (ed.), *Critical issues in qualitative research methods*, Sage Publications, Newbury Park, CA.

Leventhal, G. S., Karuza, J. J., and Fry, W. R. (1980). Beyond fairness: A theory of allocation preferences. In G. Mikula (ed.), *Justice and social interaction*, Springer-Verlag. New York, 168–218.

Lewins, F. (1993). *Writing a thesis*. Bibliotech ANUTECH Pty Ltd. Canberra, ACT.

Lightbody, J. (1995). Why study city politics? In J. Lightbody (eds.), *Canadian Metropolitics: Governing our cities*, Copp Clark, Toronto, 3–37.

Lincoln, Y., and Guba, E. (1985). *Naturalistic inquiry*. Sage Publications. Beverly Hills, CA.

Lind, E. A., and Tyler, T. R. (1988). *The social psychology of procedural justice.* Plenum Press. New York.

Lindblom, C. E. (1959). The science of muddling through. *Public Administrative Review,* 19(Spring), 79–88.

Long, R. J., and Beierle, T. C. (1999). *The federal advisory committee Act and public participation in environmental policy.* [online], 21 December 2001. http://www.rff.org.

Lynn, F. M. (1987). Citizen involvement in hazardous waste sites: Two north Carolina success stories. *Environmental Impact Assessment Review,* 7: 347–361.

Lynn, F. M., and Busenberg, G. J. (1995). Citizen advisory committees and environmental policy: What we know, what's left to discover. *Risk Analysis,* 15(2), 147–162.

Lynn, F. M., and Kartez, J. (1995). The redemption of citizen advisory committees: A perspective from critical theory. In O. Renn, T. Webler and P. Wiedemann (eds.), *Fairness and competence in citizen participation: Evaluating models for environmental discourse,* Kluwer Academic Publishers. Dordrecht, 87–101.

Maccoby, E., and Maccoby, N. (1954). The interview: A tool of social science. In G. Lindzey (ed.), *Handbook of social psychology,* Addison-Wesley, Cambridge, Massachusetts.

Marshall, C., and Rossman, G. B. (1995). *Designing qualitative research.* Sage Publications. Thousand Oaks, CA.

Mason, J. (1996). *Qualitative researching.* Sage Publications. London.

Mathbor, G. M. (1999). The perception of effective community participation: Coastal development projects in Bangladesh. PhD Thesis, Department of Social Work, The University of Calgary. Alberta.

Merton, R. K., Fiske, M., and Kendall, P. L. (1990). *The focused interview: A manual of problems and procedures.* Free Press. New York.

Mees, P. (2001). *35 up: Canberra's y-plan approaches middle age.* Royal Australian Planning Institute, National Congress 2001. Canberra.

Middendorf, D. L., and Busch, L. (1997). Inquiry for public good: Democratic participation in agricultural research. *Agriculture and Human Values,* 14: 45–57.

Minichiello, V., Aroni, R., Timewell, E., and Loris, L. (1995). *In-depth interviewing: Principles, techniques and analysis.* Longman Australia Pty Ltd. Melbourne.

Morgan, D. L. (1997). *Focus groups as qualitative research.* Sage Publications. London.

Morse, J. M., Barrett, M., Mayan, M., Olson, K., and Spiers, J. (2002). Verification strategies for establishing reliability and validity in qualitative research. *International Journal of Qualitative Methods,* 1(2), [Online] 22 March 2003, http://www.ualberta.ca/~ijqm.

Myer, C. (2002). Development plan attacked: More community consultation is needed now, local groups say. *The Belconnen Chronicle,* 22(33), 1, 20 August 2002, Canberra.

National Capital Authority (NCA), (2001). *History of the capital.* [Online], 5 May 2001. http://www.nationalcapital.gov.au/history/index.htm

Nelson, K. (1990). Common ground consensus project. In J. Crowfoot and J. Wondolleck (eds.), *Environmental disputes: Community involvement in conflict resolution,* Washington, D.C., 98–120.

Obermeyer, N. J. (1998). The evolution of public participation GIS. *Cartography and Geographic Information Systems,* 25(2), 65–66.

Octeau, C. (1999). Local community participation in the establishment of national parks: Planning for Cooperation. Master's Thesis, University of British Columbia. Vancouver.

Palerm, J. R. (2000). An empirical-theoretical analysis framework for public participation in environmental impact assessment. *Journal of Environmental Planning and Management,* 43(5), 581–600.

Pateman, C. (1970). *Participation and democratic theory.* Cambridge University Press. Cambridge.

Patton, M. (1990). *Qualitative evaluation and research methods.* Sage Publications. CA.

Patton, M. (2014). *Qualitative research and evaluation methods.* 4th Edition, Sage. Thousand Oaks, CA.

Petts, J. (2001). Evaluating the effectiveness of deliberative processes: Waste management case-studies. *Journal of Environmental Planning and Management,* 44(2), 207–226.

Pickles, J. (1995). *Ground truth: The social implications of geographical information systems.* Guilford Press. New York.

Pierce, J. C., and Doerkson, H. R. (1976). Citizen advisory committees: The impact of recruitment on representation and responsiveness. In J. C. Pierce and H. R. Doerkson (eds.), *Water politics and public involvement,* Ann Arbor Science Publishers Inc, Michigan, 249–266.

Pimbert, M. P., and Pretty, J. N. (1997). Parks, people and professionals: Putting participation into protected area management. In K. B. Ghimire and M. P. Pimbert (eds.), *Social change and conservation: Environmental politics and impacts of national parks and protected areas,* Earthscan Publications Limited, London, 297–330.

Planning ACT Together (PACTT) (2001). *Interim strategic plan for Canberra.* Canberra.

Planning and Land Management (PALM) (2000a). *The LAPAC guide.* Planning and Land Management, ACT Government. Canberra.

Planning and Land Management (PALM) (2000b). *The LAPAC protocol.* ACT Government. Canberra.

Planning and Land Management (PALM) (2001a). *High quality sustainable design outcomes: Moving towards high quality outcomes.* ACT Government. Canberra.

Planning and Land Management (PALM) (2001b). *Designing for high quality & sustainability.* ACT Government. Canberra.

Planning and Land Management (PALM) (2002a). *Establishing expectations for new neighbourhood planning.* Albert Hall., 27 February 2002, Canberra.

Planning and Land Management (PALM) (2002b). *Newsletter.* Summer 2002, ACT Government. Canberra.

Planning and Land Management (PALM) (2003a). *Community planning forums: What they do.* ACT Government. Canberra.

Planning and Land Management (PALM) (2003b). *A city like no other: your Canberra, future directions summit.* ACT Government. Canberra.

Planning and Land Management (PALM) (2003c). *LAPAC minutes, Burley griffin LAPAC, April 2003.* ACT Government. Canberra.

Planning and Land Management (PALM) (2003d). *LAPAC minutes, manuka LAPAC, April 2003.* ACT Government. Canberra.

Planning and Land Management (PALM) (2003e). *LAPAC minutes, Inner North LAPAC, May 2003.* ACT Government. Canberra.

Priscolli, J. D. (1978). *Why the federal and regional interest in public involvement in water resources development.* IWR Working Paper 78-1. U.S. Army Engineers Institute for Water Resources. Fort Belvoir, VA.

Priscolli, J. D. (1983). The citizen advisory groups as an integrative tool in regional water sources planning. In J. DeSario and S. Langton (eds.), *Citizen participation in public decision-making,* Greenwood Press, NY, 79–87.

Pross, A. P. (1992). *Group politics and public policy.* Oxford University Press. Toronto.

Ragin, C. C. (1987). *The comparative method: Moving beyond qualitative and quantitative strategies.* University of California Press. Berkeley.

Rahman, A. (1993). *People's self development: Perspectives on participatory action research*. Zed Books Ltd. London.

Rahnema, M. (1992). Participation. In W. Sachs (ed.), *The development dictionary: A guide to knowledge as power*, Zed Books Ltd, London, 116–131.

Raimond, R. R. (2001). *Trust and conflict in public participation*. Department of Public Health and Environment. Colorado.

Ramasubramanian, L. (1998). From access to power: Participatory planning and decision-making in community-based organizations using information technologies. *Regional Policy and Practice*, 7(1), 25–31.

Reid, P. (2002). *Canberra Following griffin: A design history of Australia's national capital*. National Archives of Australia. Canberra.

Reinharz, S. (1992). *Femininst methods in social research*. Oxford University Press. New York.

Renn, O., Webler, T., Rakel, H., Dienel, P., and Johnson, B. (1993). Public participation in decision-making: A three-step procedure. *Policy Science*, 26(3), 191–214.

Renn, O., Webler, T., and Weidemann, P. (1995). A need for discourse on citizen participation. In O. Renn, T. Webler and P. Weidemann (eds.), *Fairness and competence in citizen participation: Evaluating models for environmental discourse*, Kluwer Academic Publishers, Dordrecht, 1–16.

Rocha, E. M. (1997). A ladder of empowerment. *Journal of Planning Education and Research*, 17(1), 31–44.

Rosener, J. (1978a). Citizen participation: Can we measure its effectiveness? *Public Administrative Review*, 38 (September/October), 457–463.

Rosener, J. (1978b). Matching method to purpose: The challenge of planning citizen participation activities in america. In S. Langton (ed.), *Citizen participation activities*, Lexington Books, Lexington, MA.

Rosener, J. B. (1981). User oriented evaluation: A new way to view citizen participation. *Journal of Applied Behavioral Science*, 17(4), 583–597.

Ross and Associates (1991). *Lessons learned: The pacific northwest hazardous waste council's approach to regional coordination and policy development*. Ross and Associates. Washington D.C.

Rowe, G., and Frewer, L. J. (2000). Public participation methods: A framework for evaluation. *Science, Technology, & Human Values*, 25(1), 3–29.

Rubin, H. J., and Rubin, I. S. (1995). *Qualitative interviewing: The art of hearing data*. Sage Publications. Thousand Oaks, CA.

Sandelowski, M. (1986). The problem of rigor in qualitative research. *Advances in Nursing Science*, 8(3), 27–37.

Sandercock, L. (1998). *Towards cosmopolis: Planning for multicultural cities*. John Wiley & Sons. London.

Sanoff, H. (2000). *Community participation methods in design and planning*. John Wiley & Sons. NY.

Sarantakos, S. (1998). *Social research.*. Macmillan Education Australia Pty Ltd. London.

Sarkissian, W., Cook, A., and Walsh, K. (1997). *Community participation in practice: A practical guide*. Institute for Science and Technology Policy, Murdoch University. Perth.

Schatzow, S. (1977). The influence of the public on federal environmental decision-making in Canada. In W. R. Sewell and J. I. Coppock (eds.), *Public participation in planning*, Wiley & Sons, London.

Scrimgeour, D., and Hanson, L. (1993). *Advisory groups in the US department of energy environmental clean up process: A review and analysis*. Colorado Centre for Environmental Management. Denver, Colo.

Self, P. (1998). Democratic planning. In B. Gleeson and P. Hanley (eds.), *Renewing Australian planning? New challenges, new agendas*, Urban Research Progam, Australian National University, Canberra, 45–49.

Sewell, W. R. D., and Phillips, S. D. (1979). Models for the evaluation of public participation programmes. *Natural Resources Journal*, 19: 337–358.

Shapiro, S. (1986). *Scientific issues and the function of hearing procedure: Evaluating the FDA's public board of inquiry.* Duke University, Occasional. Paper, Durham, NC.Durham, NC

Shiffer, M. (1995). Interactive multimedia planning support: Moving from stand alone systems to the web. *Environment and Planning B: Planning and Design*, 22: 649–664.

Slocum, R., and Thomas-Slatyer, B. (1995). Participation, empowerment and sustainable development. In R. Slocum, L. Wichhart, D. Rocheleau and B. Thomas-Slatyer (eds.), *Power, process and participation—tools for change*, Intermediate Technology Publications Ltd, London.

Smith, L. G. (1979). An evaluative approach for the assessment of public inquiries. Master's Thesis, University of Victoria, Victoria.

Smith, L. G. (1984). Public participation in policy-making: The state-of-art in Canada. *Geoforum*, 15(2), 253–259.

Smith, L. G. (1987). The evaluation of public participation in Canada: Implications for participatory practice. *British Journal of Canadian Studies*, 2(2), 213–235.

Smith, L. G. (1993). *Impact assessment and sustainable resources management.* Longman. London.

Smith, P. D., and McDonuch, M. H. (2001). Beyond public participation: Fairness in natural resources decision making. *Society and Natural Resources*, 14: 239–249.

Soubashi, E. (1998). *Interviewing.* [Online] 2 February 2001, http://www.spinworks. demon.co.uk/pub/interview1.htm.

Stake, R. (1995). *The art of case study research.* Sage Publications. Thousand Oaks, CA.

Stein, P. (1998). 21st century challenges for urban planning: The demise of environmental planning in new South Wales. In B. Gleeson and P. Hanley (eds.), *Renewing Australian planning? New challenges, new agendas*, Urban Research Program, The Australian National University, Canberra.

Stewart, T., Dennis, R., and Ely, D. (1984). Citizen participation and judgement in policy analysis: A case study of urban air quality policy. *Policy Science*, 17(67–87).

Stewart, D. W., and Shamdasani, P. N. (1990). *Focus groups: Theory and practice.* Sage Publications. London.

Susskind, L., and Cruikshank, J. (1987). *Breaking the impasse: Consensual approaches to resolving public disputes.* Basic Books. NY.

Susskind, L., McKearnan, S., and Thomas-Larmer, J. (1999). *The consensus building handbook: A comprehensive guide to reaching agreement.* Sage Publications. Thousand Oaks, CA.

Swenson, J. D., Griswold, W. F., and Kleiber, P. B. (1992). Focus groups: Method of inquiry/intervention. *Small Group Research*, 23(4), 459–474.

Syme, G. J., and Sadler, B. S. (1994). Evaluation of public involvement in water resources planning: A researcher-practitioner dialogue. *Evaluation Review*, 18(5), 523–542.

Taberner, J., and Brunton, N. (1996). The development of public participation in environmental protection and planning law in Australia. *Environmental and Planning Law Journal*, 13, 260–268.

Talen, E. (1998). Visualizing fairness: Equity maps for planners. *Journal of the American Planning Association*, 64(1), 79–95.

Talen, E. (1999). Constructing neighborhoods from the bottom-up: The case for resident-generated GIS. *Journal of Planning Education and Research*, 31(4), 278–288.

Talen, E. (2000). Bottom-Up GIS: A new tool for individual and group expression in participatory planning. *Journal of the American Planning Association*, 66(3), 279–294.

Taylor, K. (2001a). Paving paradise to put up more houses is a bad idea. *The Canberra Times*, 31 March 2001, C3. Canberra.

Taylor, K. (2001b). Canberra needs better planning. *The Canberra Times*, 28 July 2001, C3. Canberra.

Tesch, R. (1990). *Qualitative research: Analysis types and software tools*. The Falmer Press. Hampshire.

Thibaut, J., and Walker, L. (1975). *Procedural justice: A psychological analysis*. Wiley & Sons. NY.

Thomas, J. C. (1990). Public involvement in public management: Adapting and testing a borrowed theory. *Public Administrative Review*, (July/August): 435–445.

Thomas, J. C. (1995). *Public participation in public decisions: New skills and strategies for public managers*. Jossey-Bass Publishers. San Francisco.

Tippett, J., Searle, B., Pahl-Wostl, C., and Rees, Y.. 2005. Social learning in public participation in river basin management—Early findings from HarmoniCOP European case studies. *Environmental Science amp; Policy* 8:287–299.

Troy, P. (1999). *Serving the city: The crisis in Australia's urban services*. Pluto Press Australia Limited. Sydney.

Tuler, S., and Webler, T. (1995). Process evaluation methodology. *Human Ecology Review*, 2(Winter/Spring), 62–71.

Tuler, S., and Webler, T. (1999). Voices from the forest: What participants expect of a public participation process. *Society and Natural Resources*, 12: 437–453.

Tutty, L. M., Rothery, M. A., and Grinnell, R. M. (1996). *Qualitative research for social workers*. Allyn & Bacon. Needham Heights, MA.

Tyler, T. R. (1989). The psychology of procedural justice: A test of the group-value model. *Journal of Personality and Social Psychology*, 41(4), 642–655.

Tyler, T. R., and Griffin, E. (1991). The influence of decision makers's goals on their concerns about procedural justice. *Journal of Applied Social Psychology*, 21(20), 1629–1658.

Vari, A. (1995). Citizens' advisory committee as a model for public participation: A multi-criteria evaluation. In O. Renn, T. Webler and P. Wiedemann (eds.), *Fairness and competence in citizen participation: Evaluating models for environmental discourse*, Kluwer Academic Publishers. Dordrecht, 102–115.

Vindasius, D. (1965). *Evaluation of the okanagan public involvement programme*. Water Planning and Management Branch, Environment Canada. Ottawa.

Wear, P. (1996). New age ghetto blasters. *The Bulletin*, 23–30, January: 46–48.

Webler, T. (1992). Modeling public participation as discourse: An application of Habermas's theory of communicative action. PhD Thesis, Clark University. Worcester, MA.

Webler, T. (1995). Right discourse in citizen participation: An evaluative yardstick. In O. Renn, T. Webler and P. Wiedemann (eds.), *Fairness and competence in citizen participation: Evaluating models for environmental discourse*, Kluwer Academic Publishers, Dordrecht, 35–86.

Webler, T. (1999). The craft and theory of public participation: A dialectic process. *Journal of Risk Research*, 2(1), 55–71.

Webler, T., Kastenholz, H., and Renn, O. (1995). Public participation in impacts assessment: A social learning perspective. *Environmental Impact Assessment Review*, 15(5), 443–463.

Webler, T., and Renn, O. (1995). A brief primer on participation: Philosophy and practice. In T. Webler and P. Wiedemann (eds.), *Fairness and competence in citizen participation: Evaluating models for environmental discourse*, Kluwer Academic Publishers, Dordrecht, 17–33.

Webler, T., and Tuler, S. (2000). Fairness and competence in citizen participation: Theoretical reflections from a case study. *Administration and Society*, 32(5), 566–595.

Webler, T., Tuler, S., and Krueger, R. (2001). What is a good public participation process? Five perspectives from the public. *Environmental Management*, 27(3), 435–450.

White, S. (1988). *The recent work of Jurgen Habermas : Reason, justice and modernity.* Cambridge University Press. Cambridge.

Winchester, H. P. (2000). Qualitative research and its place in human geography. In I. Hay (ed.), *Qualitative research methods in human geography*, Oxford University Press. Melbourne, 1–22.

Wolcott, H. F. (1994). *Transforming qualitative data: Description, analysis and interpretation.* Sage Publications. Thousand Oaks, CA.

World, B. (1995). *World bank participation sourcebook: towards environmentally and socially sustainable development.* Environment Department Papers, The World Bank. Washington D.C.

Yigitcanlar, T. (2000.). *A methodology for geographical information systems based participatory decision-making approach.* [online], 10 March 2000. http://likya.iyte.edu.tr/arch/city/tan/gisphd.html.

Yin, R. K. (1994). *Case study research: Design and methods.* Sage Publications. London.

Yin, R. K. (2017). *Case study research and applications: Design and methods.* Sixth Edition, Sage Publications. London.

Zeisel, J. (1984). *Inquiry by design: Tools for environment-behavior research.* Cambridge University Press. Cambridge.

Abbreviations

AAT	Administrative Appeal Tribunal
ACT	Australian Capital Authority
ACTPLA	ACT Planning and Land Authority
ALP	Austrian Labour Party
CPF	Community Planning Forum
DA	Development Applications
DRP	Design Review Panel
EPSDD	Environment, Planning and Sustainable Development Directorate
HQSD	High Quality Sustainable Design
KFDA	Kingston Foreshore Development Authority
LAPAC	Local Area Planning Advisory Committees
LDA	Land Development Authority
MLA	Member of Legislative Assembly
MPRG	Major Projects Review Group
NCA	National Capital Authority
NCP	National Capital Plan
NPG	Neighborhood Planning Group
NSW	New South Wales
PAC	Planning Advisory Committee
PACTT	Planning ACT Together
PALM	Planning and Land Management
PEA	Preliminary Environmental Assessments
TO	Technical Officers
TP	Territory Plan

Index

Printed in the United States
by Baker & Taylor Publisher Services